# 太阳能再生除湿转子性能研究

TAIYANGNENG ZAISHENG CHUSHI ZHUANZI
XINGNENG YANJIU

李 洁 孙 红 薛志超 王 展 著

中国纺织出版社有限公司

# 内 容 提 要

本书采用菲涅尔透镜聚焦太阳能辐射法再生除湿转子，利用太阳能作为再生过程的驱动力。对转子进行传热传质分析，寻找降低能量损失的有效途径。主要模拟除湿转子内能量分布、水分传递和直接光致脱附水分子动力特性。通过转子内能量分布、结构观测和除湿转轮系统性能分析实验，验证转轮内能量分布和传热传质特性，分析转轮通道流场、基材、吸附剂、再生温度、空气湿度和流速等结构参数和操作参数对除湿转轮性能和再生能量利用率等应用要素的影响。本书适合从事太阳能转化技术、太阳能除湿空调系统等相关领域的高校教师、研究生研读，也适合从事热传质、流体研究等方向的相关科研及技术人员阅读。

**图书在版编目（CIP）数据**

太阳能再生除湿转子性能研究 ＝ Performance Investigation of a Desiccant Rotor Regenerated Directly by Concentrated Solar Irradiation Using Fresnel Lens：英文／李洁等著. --北京：中国纺织出版社有限公司, 2020.10

ISBN 978-7-5180-7146-3

Ⅰ.①太… Ⅱ.①李… Ⅲ.①空调设计-英文 Ⅳ.①TB657.2

中国版本图书馆 CIP 数据核字（2019）第 281103 号

---

责任编辑：孔会云　特约编辑：陈怡晓
责任校对：楼旭红　责任印制：何　建

中国纺织出版社有限公司出版发行
地址：北京市朝阳区百子湾东里 A407 号楼　邮政编码：100124
销售电话：010—67004422　传真：010—87155801
http://www.c-textilep.com
中国纺织出版社天猫旗舰店
官方微博 http://weibo.com/2119887771
三河市宏盛印务有限公司印刷　各地新华书店经销
2020 年 10 月第 1 版第 1 次印刷
开本：710×1000　1/16　印张：6.5
字数：110 千字　定价：88.00 元

---

# Preface

With the rapid development of society, world energy consumption is expected to increase by 71% from 2003 to 2030. Energy consumption in air conditioning is estimated 45% of the whole consumption of civil and commercial buildings due to great demands of people for comfortable living environment. Meanwhile, the air conditioning industries face major problems of energy shortage and threat of global warming. Nevertheless, the traditional vapour compression system has been driven using electric power usually generated in fossil fuel power plants. Then, carbon dioxide which affects the atmospheric environment is released in large quantity.

Humidity control is obviously needed in everyday life, and it is essential in some cases e. g. hospital buildings where patients and staff members are under the risk of infection through germs, bacteria or viruses. These issues are directly associated with humidity. The vapour compression air conditioning system controls humidity by reducing the temperature below the supply air dew point, so heating is required to obtain the desired conditions of air temperature and humidity. As a result, the vapour compression air conditioning system cannot deal with the sensible and latent heat loads distinctively. However, in the desiccant air conditioning system, the latent heat load is converted into sensible heat load by means of desiccant dehumidification. Furthermore, the indoor air quality can be improved greatly by utilizing high ventilation with huge volumetric rate of the air flow. This ventilation supplies the capability of removing airborne pollutants.

For the energy crisis and environmentalfriendly problem, the desiccant air conditioning system supplies nice energy saving feature and superiority of low grade energy drive. Due to small energy consumption, the desiccant air conditioning system is a massive energy saver, and it is one of effective way to relieve energy conservation issues. The desiccant air conditioning system can utilize many kinds of available driving heat sources; solar energy, geothermal energy, exhausts heat from power station and so on.

Besides, there is no environmental pollution. On the other hand, heat pollution can be removed from the factory. The conventional vapour compression air conditioning system poses global environmental issues of ozone layer depletion, greenhouse gas effect or global warming, etc. , whereas the desiccant air condition system is regarded as a green technology which ensures environmental safety. According to the recent study on wet markets of Hong Kong, the desiccant air conditioning system achieves $1\% \sim 13\%$ reductions in $CO_2$ emission as compared to the vapour compression air conditioning system.

Consequently, it is imperative to develop the high-performance desiccant cooling technologies to substitute the conventional vapour compression systems.

This book is divided into five chapters:

Chapter 1    Introduction

Chapter 2    Measurements of energy distribution within narrow channels

Chapter 3    Predictions for energy distribution

Chapter 4    Experiment for measuring water vapour adsorption/desorption rate in desiccant rotor

Chapter 5 Simulation of dehumidifying process of desiccant rotor

This research was supported by the National Natural Science Foundation of China (51776131, 51906166, 51805337) and Natural Science Foundation of Liaoning Province (lnzd201901).

Participants in the preparation of this book include: Li Jie (Overall responsibility), Sun Hong (Proofreading), Wang Zhan(Proofreading)from School of Mechanical Engineering,and Xue Zhichao (Proofreading) from School of science, Shenyang Jianzhu University.

Thanks to professor Hideo Mori and Associate Professor Yoshinori Hamamoto of Kyushu University for their guidance on experiment and simulation work and for their valuable Suggestions on the content of this book. Readers are invited to give their valuable comments on the inadequacies of the book.

# Contents

# Chapter 1   Introduction

## 1.1   Introduction

### 1.1.1   Background

The loads of air conditioning are separated into two types: sensible heat load (thermal load) and latent heat load (moisture load). For the air conditioning system improving an indoor air condition with fresh outside air, the latent heat load is an indispensable component of the entire air conditioning load. In the traditional vapour compression air conditioning system, the process air, that is, fresh air is cooled below the dew-point temperature for condensation dehumidifying. Then, both the sensible and latent heats are dealt with. However, to control the indoor air humidity effectively, it is preferable to handle the sensible heat and latent heat independently.

A desiccant air conditioning system based on adsorption effect of adsorbents for water vapour dehumidifying has been developed and improved. The latent heat load and sensible heat load can be handled separately using a desiccant cooling system. Therefore, the desiccant air conditioning system has attracted great attention due to the demands for high quality living environment[1].

With the rapid development of society, the world energy consumption is expected to increase by 71% from 2003 to 2030[2]. Energy consumption in air conditioning is estimated 45% of the whole consumption of civil and commercial buildings due to great demands of people for comfortable living environment[3]. Meanwhile, the air conditioning industries face major problems of energy shortage and threat of global warming. Nevertheless, the traditional vapour compression system has been driven using electric power usually generated in fossil fuel power plants. Then, carbon dioxide which affects the atmospheric environment is released in large quantity.

Consequently, it is imperative to develop the high-performance desiccant cooling technologies to substitute the conventional vapour compression systems[4].

## 1.1.2 Overview of desiccant cooling system

The desiccant cooling system generally includes three components: a desiccant dehumidifier with effective desiccant material, a driving thermal resource for regeneration of desiccant material, and a cooling device reducing the temperature of dehumidified air. The principle of the desiccant cooling system is shown in Fig. 1. 1[5]. In the case where a solid state desiccant material is employed, a rotating desiccant rotor or a periodically regenerated adsorbent bed is generally used as the desiccant dehumidifier. The process air, that is humid ambient air, is dehumidified during passing through the desiccant dehumidifier, and it becomes dry air due to being removed the water vapour. Then, the dry air is supplied into the indoor room after cooling process which corresponds to sensible heat control.

The key component of the desiccant cooling system is the desiccant dehumidifier. Whether the process of dehumidifying is adsorption using solid desiccant or absorption by liquid desiccant, the principle is nearly identical, as follows.

The incoming fresh airflow is forced to pass through the desiccant material in the dehumidifier. Then, the amount of water vapour in airflow is adsorbed or absorbed effectively to satisfy the desired indoor humidity condition. To operate the desiccant cooling system continually, regeneration of the desiccant material is required; driving out of the adsorbed or absorbed water vapour from the desiccant material. To adsorb or absorb the water vapour favourably in the next step, the desiccant material must be dried sufficiently. So, this regeneration process of the desiccant material is a fatal link to operate the desiccant cooling system effectively, and it is achieved by heating the desiccant material up to its regeneration temperature. The regeneration temperature depends on material of the desiccant. Selection of the material in the desiccant cooling system must be made with careful consideration. Fig. 1. 1 shows the principle of desiccant dehumidification and air conditioning.

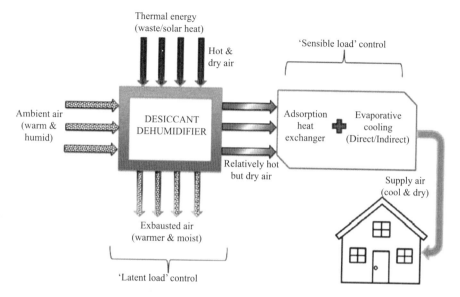

Fig. 1. 1   Principle of desiccant dehumidification and air conditioning[5]

All of desiccant cooling systems belong to open cycles. From the types of desiccant, desiccant air condition systems could be divided into two categories: solid adsorption system and liquid absorption system. The former has two types; adsorbent bed type and rotating wheel type. The adsorbent bed type of adsorptive dehumidification apparatus achieves the intermittent moisture adsorption regeneration by changing the process airflow and hot airflow alternately. And the rotating wheel type can realize continuous dehumidification, that is, regeneration by rotating the desiccant wheel across the two flows of process air and hot air. Thereby the rotating wheel type has been already applied widely in the desiccant air conditioning system. When the liquid desiccant is employed, the dehumidifier (absorber) is the equipment inside which the liquid desiccant is brought into to contact with the process air stream. The dehumidifier (absorber) and the regenerator are generally referred to as contactors.

### 1.1.3 Advantages of desiccant cooling system

The traditional vapour compression air condition system takes a great proportion of building energy consumption. Although according to incomplete statistics, conventional air conditioning makes up over half of the electric energy consumption of the building. As a result of condensation dehumidification, working system usually burns more fossil fuels to meet the requirements of handling sensible and latent heat loads at the same time. For the energy crisis and environmental-friendly problem, the desiccant air conditioning system shows nice energy saving feature and superiority of low grade energy drive.

Humidity control is obviously needed in everyday life, and it is essential in some cases e. g. hospital buildings where patients and staff members are under the risk of in fection through germs, bacteria or viruses. These isues are directly associated with humidity[6]. The vapour compression air conditioning system controls humidity by reducing the temperature below the supply air dew-point, so heating is required to obtain the desired conditions of air temperature and humidity[7-10]. As a result, the vapour compression air conditioning system cannot deal with the sensible and latent heat loads distinctively. However, in the desiccant air conditioning system, the latent heat load is converted into sensible heat load by means of desiccant dehumidification. Furthermore, the indoor air quality can be improved greatly by utilizing high ventilation with huge volumetric rate of the air flow. This ventilation supplies the capability of removing airborne pollutants.

Due to small energy consumption, the desiccant air conditioning system is a massive energy saver, and it is one of effective ways to relieve energy conservation issues. The desiccant air conditioning system can utilize many kinds of available driving heat sources; solar energy, geothermal energy, exhausts heat from power station and so on. Besides, there is no environmental pollution. And heat pollution can be removed from the factory[11-13]. The conventional vapour compression air conditioning system poses global environmental issues of ozone layer depletion, greenhouse gas effect or global warming, etc.[14], whereas the desiccant air condition system is regarded as a green

4

technology which ensures environmental safety[15]. According to a recent study on wet markets of Hong Kong, the desiccant air conditioning system achieves 1% ~ 13% reductions in $CO_2$ emission as compared to the vapour compression air conditioning system[16]. Furthermore, the desiccant air conditioning system yields 24% of electricity saving of in hot and humid climate of Thailand[17].

## 1.2　Solid desiccant cooling research

### 1.2.1　Basic structure and principle

Fig. 1.2 presents the fundamental running principle of rotary desiccant dehumidifier schematically. The desiccant material is put into bracing structure of supporting material. The circumferential cross section of wheel is divided into two parts, in the form of a clapboard; process air part and regeneration air part[18]. When the wheel rotates through two separate sections constantly, the process air becomes dried by adsorption effort of the desiccant material. Meanwhile, rich water vapour in the desiccant material is desorbed by the regeneration air heated by a heater in the regeneration air part. Then, the regeneration air is humidified. By the wheel rotation, the adsorption and desorption (regeneration) processes are repeated one after another in cycles, Only thing to note here is that the desiccant dehumidification (desorption) process is similar to an isenthalpic procedure, namely, it merely has a process of energy conversion from latent heat to sensible heat and there by produces useless cooling. Therefore, to accommodate cooling influence, assistant cooler, like evaporation cooler and other air condition apparatus must be incorporated to eliminate the sensible heat. The performance of desiccant air conditioning systems is principally determined by the system configuration when the desiccant material, wheel structure and operation condition are given. Because of this, various rotary desiccant air conditioning systems have been proposed and studied based on the oretical analysis and experiments[5, 19].

5

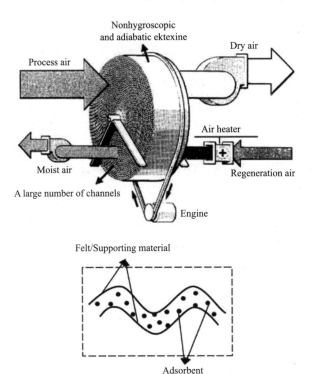

Fig. 1. 2　Structure of rotary desiccant dehumidifier schematically[18]

## 1.2.2　Historical development and progress of solid desiccant cooling system

In 1955, Pennington[20] was awarded the first patent of desiccant cooling system and proposed Pennington cycle as shown in Fig. 1. 3 After then, researchers from all over the world have worked for desiccant systems actively, and constructed various air conditioning systems. The institute of gas technology in the United States of America developed a molecular sieve as the desiccant in the adsorption apparatus.

Pennington cycle, known as ventilation cycle, is shown in Fig. 1. 3 schematically. Ambient air is adopted as the process air at state point 1 and goes through a desiccant wheel, where moisture is removed from the process air and its temperature is increased for the adsorption heat effect. Continuously, the hot dried air is cooled from state point 2 to point 3 through a heat exchanger. Thereafter, the process air is cooled to the supply

6

air state by passing through a direct evaporative cooler. On the regeneration air side, the return air is cooled and humidified in a direct evaporation cooling device from state point 5 to point 6. This air is then apparently heat-exchanged with the process air to pre-cool the process air and to pre-heat itself. The warm air stream is then further heated from state point 7 to point 8 by a heat source. After regenerating the desiccant wheel, the air is exhausted at state point 9.

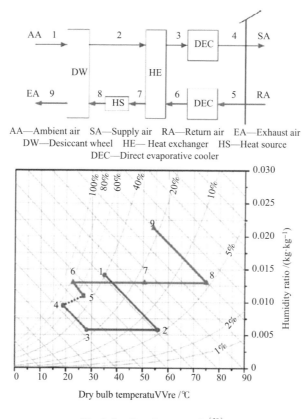

Fig. 1.3   Pennington cycle[20]

In the 1960's, Dunkle proposed a fundamental cycle of desiccant air condition. Dunkle cycle combines the merits of ventilation cycle, which can provide cold air at relatively low temperature, and recirculation cycle[21]. This cycle requires a large amount of airflow circulation. Collier and Cohenproposed the staged regeneration idea[22]. In the first stage, part of the airflow without heating is delivered to dryer for regenerating the desiccant. Second stage, another part of airflow goes through the dryer and heated by it, reaching regeneration temperature, and then works for the desiccant. The results showed it improved the cooling capacity. Schultz[23] used the solar radiation for the desiccant regeneration directly. The results showed COP of that system was lower than that of using air.

In addition, there are a lot of studies about the combination of rotating wheel system and traditional air condition system. Babus and Probert[24] compared the economy of many rotating wheel systems which had internal combustion driving with natural gas. The waste heat generated by the internal combustion engine provided the energy needed by regeneration the rotating wheel. Simulation results showed that under all the conditions, running cost was lower than electronic driving system. Kiatsiriroat[25] combined silica gel rotating wheel system with heat pump system and gave the theoretical simulation and experiment study. The results presented, when the inlet airflow temperature was between 30 ~ 50℃, the desiccant device can reduce the load of heat pump by 7% ~ 20%. Dai[26] proposed a new kind of composite system of solid desiccant device with solid absorption cooling. Compared with traditional refrigeration systems, the new system achieved heat property coefficient bigger than 0.4 and the outlet airflow temperature lower than 20℃. Parsons[27] studied a desiccant subsystem combined with vapour compression system. This composite system had desiccant rotating wheel with parallel channel. It turned out that the COP of vapour compression system was enhanced by 40% with the help of desiccant subsystem. Khalid[28] combined the desiccant rotating wheel with conventional air compression system and compared with traditional air condition system by simulations and experiments. Results indicated that the compound system can save the energy by 40%. Babus and Probert[24] also gave a proposal about the combination of solid rotating wheel air condition and co-generation of heat and electricity using high temperature exhaust to drive the rotating wheel regeneration.

Furthermore, there still area certain amount of theoretical researches on rotor air condition, including the following aspects; Building the mathematical model to predict the performance of desiccant air conditioning by combining different kinds of models[29-30], analyses of the performance changing under various climates[31], optimizing

8

desiccant system running parameters, and some successful examples of experiments[32].
Over the last 20 years, there were lots of studies on rotating wheel desiccant air conditioning using thermodynamics second law [33-35].

## 1.2.3   Energy source for desiccant rotor regeneration

For a few years, the rotary desiccant air conditioning system has experienced an advanced commercial application growing. Since 1980's, the rotary desiccant dehumidification has been well-developed on industrial market. Generally, in practical application, rotary desiccant air conditioning systems are divided into two categories; a solar-powered rotary desiccant air conditioning system and a rotary desiccant air conditioning system powered by other low grade heatsources, such as district heat, heat supplied from a combined heat and power plant, waste heat and bio-energy. Enhancing the utilization of low grade energy is a great challenge nowadays. Researchers focusing on solar energy become more and more. The solar thermal system which is very useful in meeting the cooling demand perfectly is expected to support a large part of the energy required by air conditioning. Therefore, interest in coupling the rotary desiccant air conditioning and solar energy was boosted up. Many experimental projects have been conducted[3, 31].

A solar desiccant cooling system equipped with flat plate solar collectors of 22.5 $m^2$ and a radiant ceiling of 76.1 $m^2$ was set up at University of Palermo, Italy, as shown in Fig. 1.4[36]. A hot water tank and an auxiliary boiler were attached in this application. Back and fourth of the desiccant wheel, two cooling coils are set up to precool and pre-dehumidify the process air, when the heat recovery cooling capacity is not enough. This solar desiccant air conditioning system took a peak summer load of 28.8 kW with thermal COP of 0.86. Another solar desiccant cooling system with hybrid photovoltaic (PV)/thermal collector has been developed recently in Orbassano, Italy[37]. The heat recovered from the hybrid and the cogeneration was used for dehumidifying the renewal air in summer, whereas the heat was used for heating in winter.

Two types of two-stagesystem have been developed by some researchers in Shanghai Jiao Tong University of China[38], namely, two-stage rotary desiccant cooling systemusing two desiccant wheels and one-rotor two-stage rotary desiccant cooling

Fig. 1. 4    Solar desiccant cooling system installed at University of Palermo, Italy[36]

system based on one wheel[39]. Climate of the southeast China is hot and humid in summer. Deep dehumidification was achieved without high regeneration temperature and big
initial investment in solar collectors.

Two-stage rotary desiccant cooling system placed in Jiangsu, China is shown in
Fig. 1. 5. An air-source vapour compression air conditioning unit is used to make sure the
operating continuity under cloudy and rainy condition. Flat plate solar collectors are also
needed to ensure good integration into building. Under typical environment, the
solar-powered desiccant cooling unit showed good energy saving, achieving an average
cooling capacity above 10 kW. The system could convert more than 40% of the received
solar radiation to the capability of air conditioning in fine days. One-rotor two-stage rotary desiccant cooling system was installed in Shanghai Jiao Tong University[39]. Solar

air collectors of 15 m$^2$ were used to generate hot air. In summer, the solar heated air was introduced into the unitto regenerate the desiccant wheel. While, in winter, the system worked in two kinds of modes, direct solar heating mode and solar heating with desiccant humidification mode. Under solar heating mode, fresh air over 30℃ was provided. The solar heating with desiccant humidification mode was developed owing to simultaneous heating and humidifying of the process air and conditioning the space to a better state.

(a)

(b)

Fig. 1. 5　Two-stage solar desiccant cooling system installed in Jiangsu, China[38]

Generally, solar energy depends on weather conditions and geometrical position greatly. Even in the area with abundant solar radiation, the performance of solar collector would be poor and could not totally meet the heat requirement of thermal pow-ered unit in cloudy or rainy weather. For these reasons, most of the installed desiccant cooling systems are coupled with gas burner, district heating or heat from a combined heat and power plant, etc[40-44]. Todate, wide applications have arisen, especially in locations with high moisture production or requiring tight control of moisture level; such as supermarket, ice arena, theatre, store house and hospital operating room[40,42].

## 1.3  Analysis methods for desiccant dehumidification

### 1.3.1  Theoretical studies

With the development of computer hardware and numerical methodology, mathe-matical models are being used to carry out critical investigations concerning on the desic-cant wheel. This method has many advantages. It takes less time and cost than experi-mental method for predicting the performance of a desiccant wheel. The second point is that the mathematical model can produce large volumes of results at virtually no added expense and it is very convenient to perform parametric research and optimization analy-sis. The third is that it is possible to determine values of some parameters which cannot be measured by experimental means. Finally, the models help to investigate a novel de-vice applying desiccant dehumidifying process because the fundamental physics and sur-face chemistry of adsorption phenomena considered in the models are common [45-48].

Lots of mathematical models have been constructed and employed to analyse, de-velop and design desiccant wheels in the early years. The modelling of the enthalpy re-covery wheel was based on the analogy between heat and mass transfer, in early 1970's. For a limitation of the computational power, analogy between heat and mass transfer was used because the methods for heat transfer are readily available. Thereby, this had no significance with the fast development by computers and technologies[49].

Simonson and Besant[50-51] developed a one dimensional transient model to give a simulation on the heat and moisture transfer in the wheel. Reasonable agreement between the experiment and the model predictions was achieved by using experimental data. Van den Bulck and Mitchell et al. [52-53] made an Effectiveness – Number of Transfer Units method for the active desiccant wheel. This model was used to explore the influences of different running variables, such as the regeneration air flowrate, regeneration temperature and wheel rotary speed, so as to maximize the wheel COP based on the thermal and electrical energy input.

The mentioned above models only included the convective heat and mass transfer resistances on the gas solid interface. In addition to the resistance at the interface, the solid side resistances for heat and mass transfer were also considered in Majumdar's model[54]. The moisture transport in the desiccant matrix was represented by gas diffusion and surface diffusion resistances. Zhang, Niu[55], Sphaier and Worek[56] studied two dimensional models considering both heat and mass transfer resistances in both axial and thickness directions of the solid desiccant. Their models gave some insights into the heat and mass transfer processes in the desiccant matrix, which was never mentioned in the previous publications.

In the last few years, the passion of researchers for theoretical studies with solid desiccants keeps moving. The simulation optimization of solar−assisted desiccant cooling system for subtropical Hong Kong was published by Fong et al[59]. After optimization, the monthly average solar fraction was in the range from 8% to 33%, and the yearly average was 17%. Panaras et al. [58] gave an analysis on a solid desiccant air−conditioning system for recognizing the main design parameters, and examined their effects on the performance of the system. Panaras et al. [59] developed and experimentally validated a model for the operation of a desiccantair−conditioning system. Finocchiaro et al. [60] developed a theoretical analysis on an advanced solar assisted desiccant and evaporative cooling system equipped with wet heat exchangers. Ali[61] analysed and showed that it was feasible to implement desiccant enhanced nocturnal radiative cooling solar collector system for air comfort application in hot dry areas of Upper Egypt.

### 1.3.2 Experimental studies

By experiments, a number of studies have demonstrated the potential for harnessing solar energy to drive the desiccant. In the current situation, most researchers focus on using solar collector to convert the solar energy into hot water or hot air.

The use of compound desiccant was proposed by Jia et al. [62] to develop high performance desiccant cooling system. Experimental results presented that, under a practical operation, the novel desiccant wheel can remove more moisture from the process air by about 20% ~40% over the desiccant wheel employing regular silica gel. The construction and initial operation of a combined solar thermal and electric desiccant cooling system was reported by Enteria et al[63]. This system could run during both nighttime and daytime. The nighttime operation used the auxiliary electric heater for thermal energy storage, while the daytime operation was driven on solar energy collection and desiccant cooling. An experimental investigation of a solar desiccant cooling installation powered by vacuum-tube solar collectors was carried out by Bourdoukan et al[64]. Also, Eicker and others[65] reported the operational results with solar air collector driving desiccant cooling systems installed in Spain, Germany and China. The development and construction of a novel solar cooling and heating system was presented by Enteria et al [66]. They reported that the thermal energy subsystem functioned well with expected performance in solar energy collection and thermal storage. An experimental investigation was reported about a one-rotor two-stage desiccant cooling/heating system driven by solar air collectors[67]. Experimental results proved that the solar COP of its system was about 0.45 when thermal efficiency of collector was 50%. Another experimental analysis was carried out on the dehumidification and thermal performance of a desiccant wheel by Angrisani et al[68]. The results indicated that, compared to the process air temperature, the process air humidity ratio and regeneration temperature had a bigger impact on the desiccant wheel performance. For an institutional building in subtropical Queensland, Australia, the analysis of solar desiccant cooling system was presented by Baniyounes et al[69]. The installed cooling system consisted of $10m^2$ solar collectors and a $0.4m^3$ hot water storage tank, and the estimated annual primary energy saving was 22%.

On the whole, at present, a great number of researches have concentrated on how to improve the thermodynamic coefficient of performance and energy usage ratio of rotary desiccant system. Actually, the main disadvantage of rotary desiccant air condition is that the temperature of the desiccant rises during the adsorption process due to the release of the adsorption heat, and it reduces the ability of moisture absorption. Also, there is large loss of heat and mass transfer in the whole process. Besides, the high regeneration temperature prevents realizing the ideal constant temperature desiccant process. According to the statistics, in the middle level of hot and humid condition, the required regeneration temperature of the rotary desiccant air conditioning is more than 80℃. For the extreme high hot and humid case, the value is even more than 95℃. The existing heating method uses solar collectors generally. If we want to provide so high temperature regeneration air, there must be some new methods to use the solar energy effectively.

# References

［1］ ALBERS J, KÜHN A, PETERSEN S, et al. Development and progress in solar cooling technologies with sorption systems［J］. Chemie Ingenieur Technik, 2011, 1853-1863.

［2］ WANG Y H, RADERMACHER R, ALILI A A, et al. Review of solar cooling technologies［J］. HVAC&R Res, 2008, 3(14):507-528.

［3］ HENNING H M, Solar assisted air conditioning of buildings – an overview［J］. Applied Therm Engineering, 2007, 27(10):1734-1749.

［4］ ZHENG X, GE T S, WANG R Z. Recent progress on desiccant materials for solid desiccant cooling systems［J］. Energy, 2014,74: 280-294.

［5］ SULTAN M, EL-SHARKAWY I I, MIYAZAKI T, et al. An overview of solid desiccant dehumidification and air conditioning systems［J］. Renewable and Sustainable Energy Reviews, 2015,46:16-29.

［6］ SOOKCHAIYA T, MONYAKUL V, THEPA S. Assessment of the thermal environment effects on human comfort and health for the development of novel air conditioning system in tropical regions［J］. Energy Build, 2010,10(42):1692-1702.

［7］ JEONG J, YAMAGUCHI S, SAITO K, et al. Performance analysis of four-partition desiccant wheel and hybrid dehumidification air-conditioning system［J］.

Int. J. Refrig, 2010,3(33):496-509.

[8] SUN D W, Solar powered combined ejector-vapour compression cycle for air conditioning and refrigeration[J]. Energy Convers Mgmt, 1997,5(38):479-491.

[9] KHALID C S, A, GANDHIDASAN P, FARAYEDHI A, Simulation of a hybrid liquid desiccant based air-conditioning system[J]. Appl. Therm. Eng., 1997, 2(17):125-134.

[10] GOMMED K, GROSSMAN G. Experimental investigation of a liquid desiccant system for solar cooling and dehumidification[J]. Sol. Energy, 2007,1(81): 131-138.

[11] HÜRDOĞAN E, BÜYÜKALACA O, YLMAZ T, et al. Investigation of solar energy utilization in a novel desiccant based air conditioning system[J]. Energy Build, 2012,55, 757-764.

[12] HALLIDAY S P, BEGGS C B, SLEIGH P A. The use of solar desiccant cooling in the UK: a feasibility study[J]. Appl. Therm. Eng., 2002,12(22): 1327-1338.

[13] MAVROUDAKI P, BEGGS C B, SLEIGH P A, et al. The potential for solar powered single-stage desiccant cooling in southern Europe[J]. Appl. Therm. Eng., 2002,10(22): 1129-1140.

[14] HEIDARINEJAD G, HEIDARINEJAD M, DELFANI S, et al. Feasibility of using various kinds of cooling systems in a multi-climates country[J]. Energy Build., 2008,40: 1946-1953.

[15] ZHANG L Z, FU H X, YANG Q R, et al. Performance comparisons of honeycomb-type adsorbent beds[J]. Energy, 2014,65:430-440.

[16] LEE S H, LEE W L. Site verification and modeling of desiccant-based system as an alternative to conventional air-conditioning systems for wet markets[J]. Energy, 2013,55,1076-1083.

[17] HIRUNLABH J, CHAROENWAT R, KHEDARI J, et al. Feasibility study of desiccant air-conditioning system in Thailand[J]. Build. Environ, 2007,2 (42),572-577.

[18] GE T S, LI Y, WANG R Z, et al. A review of the mathematical models for pre-

dicting rotary desiccant wheel[J]. Renew Sustain. Energy Rev, 2008,6(12): 1485-1528.

[19]    HSU S T, LAVAN Z, WOREK W M. Optimization of wet-surface heat exchangers[J] . Energy, 1989,11(4):757-770.

[20]    PENNINGTON N A. Humidity changer for air conditioning[J]. Patent No. 2, 1955,537-700.

[21]    LA D, DAI Y J, LI Y, et al. Technical development of rotary desiccant dehumidification and air conditioning: A review [J]. Renew. Sustain. Energy Rev. , 2010,1(14): 130-147.

[22]    COLLIER R K, COHEN B M. An analytical examination of methods for improvingthe performance of desiccant cooling systems [J]. J. Sol. Energy Eng, 1991, 113, 157-163.

[23]    SCHULTZ K J, Rotary solid desiccant dehumidifiers analysis of models and experimental investigation[D]. University of Wisconsin, Madison, USA, 1987.

[24]    BABUS HAQ R F, OLSEN H, PROBERTS D. Feasibility of using an integrated small-scale CHP unit plus desiccant wheel in a leisure complex[J]. Appl Energy, 1996,1(53):179-192.

[25]    KIATSIRIROAT T, TACHAJAPONG W. Analysis of a heat pump with solid desiccant tube bank[J]. Int. J. Energy Res, 2002,2(26):527-542.

[26]    DAI Y J, WANG R Z, XU Y X, Study of a solar powered solid adsorption - desiccant cooling system used for grain storage[J], Renew. Energy, 2002,3 (25):417-430.

[27]    PARSONS B K, PESARAN A A, Improving Gas-Fired Beat Pump Capacity and Performance by Adding a Desiccant Dehumidification Subsystem[C]. Proc of the Seventh Symposium on Improving Building Systems in Hot and Humid Climates, 1990.

[28]    KHALID A, MAHMOOD M, ASIF M, et al. Solar assisted, pre-cooled hybrid desiccant cooling system for Pakistan [J]. Renew. Energy, 2009, 1 (34): 151-157.

[29]    RUIVO C R, COSTA J J, FIGUEIREDO R. On the behaviour of hygroscopic

wheels: Part I – channel modelling[J]. Int. J. Heat Mass Transf, 2007, 50, 4812-4822.

[30] NELSON J S, BECKMAN W, MITCHELL J W, et al. Simulations of the performance of open cycle desiccant systems using solar energy[J]. Sol. Energy, 1978, 21, 273-278.

[31] HENNING H M, ERPENBECK T, HINDENBURG C, et al. Otential of solar energy use in desiccant cooling cycles[J]. Int. J. Refrig. , 2001, 3 ( 24 ): 220-229.

[32] JIN W, KODAMA A, GOTO M, An adsorptive desiccant cooling using honeycomb rotor dehumidifier[J]. Chem Eng. , 1998, 5(31): 706-713.

[33] SHEN C M, WOREK W M. The second-law analysis of a recirculation cycle desiccant cooling system: Cosorption of water vapor and carbon dioxide[J]. Atmos. Environ. , 1996, 9(30): 1429-1435.

[34] PONS M, KODAMA A. Entropic analysis of adsorption open cycles for air conditioning. Part 1: first and second law analyses[J]. Int. J. Energy Res, 2000, 24, 251-262.

[35] KODAMA A, JIN W, GOTO M, et al. Entropic analysis of adsorption open cycles for air conditioning. Part 2: interpretation of experimental data[J]. Int. J. Energy Res, 2000, 263-278.

[36] SOLAR Heating and Cooling of Buildings, eco buildings Guidelines 2007. BRITA inPuBs. , The 6th Framework Programme of the European Union, http://www. brita-in-pubs. eu/bit/uk/03viewer/retrofit_measures/pdf/FINAL_11_SolarCooling_Marco_01_4_08b. pdf

[37] BUTERA F, ADHIKARI RS, ASTE N, et al. Hybrid photovoltaic-thermaltechnology and solar cooling: the CRF solar facade case study[J/OL].. http://www. pvdatabase. org/pdf/FIAT-SolarFacade. pdf

[38] GE T S, LI Y, WANG R Z, et al. Experimental study on a two-stage rotarydesiccant cooling system[J]. Int J Refrig, 2009, 3(32):498-508.

[39] LI H, DAI Y J, LI Y, et al. Experimental investigation on a one – rotor two-stage desiccant cooling/heating system driven by solar aircollectors[J].

Appl Therm Eng. , 2011,31, 3677-3683.

[40]    MAZZEI P, MINICHIELLO F, PALMA D, HVAC dehumidification systems for thermal comfort: A critical review[J]. Appl. Therm. Eng, 2005,5(25): 677-707.

[41]    CASAS W, SCHMITZ G, Experiences with a gas driven, desiccant assisted air conditioning system with geothermal energy for an office building[J]. Energy Build, 2005,5(37):493-501.

[42]    MAZZEI P, MINICHIELLO F, PALMA D, Desiccant HVAC systems for commercial buildings[J]. Appl. Therm. Eng, 2002.5(22): 545-560.

[43]    SPEARS J W, JUDGE J. Gas-fired desiccant system for retail super center [J]. ASHRAE J, 1997,39, no. 10.

[44]    MILLER J, The Performance of a Desiccant-Based air Conditioner on a Florida School[J] . ASHRAE Trans, 2001,1(108): 575-586.

[45]    KLEIN H, KLEIN S, MITCHELL J W. Analysis of regenerative enthalpy exchangers[J]. Int. J. Heat Mass Transf, 1990,4(33): 735-744.

[46]    SAN J Y. Heat and mass transfer in a two- dimensional regenerator with a solid conduction effect[J]. 1993, 3(36): 1993.

[47]    MAJUMDAR P, WOREK W. Combined heat and mass transfer in a porous adsorbent[J]. Energy, 1989,3(14): 161-175.

[48]    MAJUMDAR P, WOREK W M . Combined heat and mass transfer in a porous adsorbent[J]. energy, 1989, 14(3):161-175.

[49]    SIMONSON C J. Heat and moisture transfer in energy wheels[D]. Canada, University of Saskatchewan, 1987.

[50]    SIMONSON C J, BESANT R W. Heat and moisture transfer in desiccant coated rotary energy exchangers: part I . numerical model heat and moisture transfer in desiccant coated rotary energy exchangers: part I . numerical model [ J ]. HVAC&R Res,1997, 3(4): 325-350.

[51]    SIMONSON C. J,BESANT R W. Heat and Moisture Transfer in Desiccant Coated Rotary Energy Exchangers: Part II. Validation and Sensitivity Studies[J]. HVAC&R Res. ,1997,3(4):325-350.

[52] BULCK E V D, MITCHELL J W, KLEIN S A. Design theory for rotary heat and mass exchangers - I. Wave analysis of rotary heat and mass exchangers with infinite transfer coefficients[J]. Heat Mass Transf,1985,8(28):1575-1586.

[53] Bulck E V D, Mitchell J W, Klein S A. Design theory for rotary heat and mass exchangers - II. Effectiveness-number-of-transfer-units method for rotary heat and mass exchangers[J]. Heat Mass Transf. ,1985,28(8):1587-1595.

[54] MAJUMDAR P. Heat and mass transfer in composite desiccant pore structures for dehumidification[J]. Sol Energy, 1998, 1(62): 1-10.

[55] ZHANG L Z, NIU J L. Performance comparisons of desiccant wheels for air dehumidification and enthalpy recovery [J]. Appl. Therm. Eng. , 2002, 22, 1347-1367.

[56] MAJUMDAR P . Heat and mass transfer in composite desiccant pore structures for dehumidification[J]. Solar Energy, 1998, 62(1):1-10.

[57] ZHANG L. Z, LIU J L. Performance comparisons of desiccant wheels for air dehumidification and enthalpy recovery[J]. Applied Thermal Engineering, 2002.

[58] SPHAIER L A, WOREK W M. Analysis of heat and mass transfer in porous sorbents used in rotary regenerators [J]. Int J Heat Mass Transf, 2004, 47, 3415-3430.

[59] FONG K F, CHOW T T, LIN Z, et al. Simulation-optimization of solar - assisted desiccant cooling system for subtropical Hong Kong[J]. Appl Therm Eng,2010, 30,220-228.

[60] PANARAS G, MATHIOULAKIS E, BELESSIOTIS V. Solid desiccant air-conditioning systems - Design parameters[J]. Energy, 2011,5(36):2399-2406.

[61] PANARAS G, MATHIOULAKIS E, BELESSIOTIS V et al. Theoretical and experimental investigation of the performance of a desiccant air - conditioning system[J]. Renew Energy, 2010,7(35):1368-1375.

[62] FINOCCHIARO P, BECCALI M, NOCKE B. Advanced solar assisted desiccant and evaporative cooling system equipped with wet heat exchangers[J]. Sol Energy, 2012,1(86): 608-618.

[63] ALI A H H. Desiccant enhanced nocturnal radiative cooling-solar collector sys-

tem for air comfort application in hot arid areas[J]. Sustain Energy Technol. Assessments, 2013,1,54−62.

[64] JIA C X, DAI Y J, WU J Y, et al. Use of compound desiccant to develop high performance desiccant cooling system[J]. Int J Refrig, 2007,2(30):345−353.

[65] ENTERIA N, YOSHINO H, MOCHIDA A, et al. Construction and initial operation of the combined solar thermal and electric desiccant cooling system[J]. Sol Energy, 2009,8(83):1300−1311.

[66] BOURDOUKAN P, WURTZ E, JOUBERT P. Experimental investigation of a solar desiccant cooling installation[J]. Sol Energy, 2009, 11(83):2059−2073.

[67] EICKER U, SCHNEIDER D, SCHUMACHER J, et al. Operational experiences with solar air collector driven desiccant cooling systems[J]. Appl. Energy,2010, 12(87):3735−3747.

[68] ANGRISANI G, MINICHIELLO F, ROSELLI C, et al. Experimental analysis on the dehumidification and thermal performance of a desiccant wheel[J]. Appl. Energy, 2012,92, 563−572.

[69] BANIYOUNES A M, LIU G, RASUL M G, et al. Analysis of solar desiccant cooling system for an institutional building in subtropical Queensland, Australia [J]. Renew Sustain Energy Rev, 2012,8(16):6423−6431.

# Chapter 2 Measurements of energy distribution within narrow channels

In this chapter, the results of measurement of energy distribution within narrow channels are described. The purpose of this measurement is to find the distribution characteristics of absorbed irradiation energy rate on the narrow channel wall of desiccant rotor with direct heating.

## 2.1 Energy distribution measurement of laser irradiation

### 2.1.1 Concept of measurement

The principle of the measurement is shown in Fig. 2.1. In the desorption process of the direct heating desiccant rotor system, after concentrating by Fresnel lens, the solar irradiation arrives in the front of the desiccant rotor from different incident angles. Some of the solar irradiation can be absorbed by the adsorbent directly. Others enter narrow channels and pass through the channel to the outlet side in the air flow direction. The incident angle of input solar irradiation increases as the radius of desiccant rotor

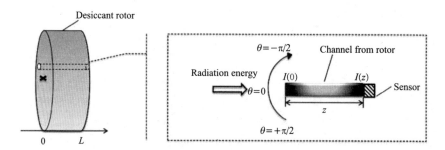

Fig. 2.1　Concept of energy distribution measurement

increases as shown in Fig. 2. 2.

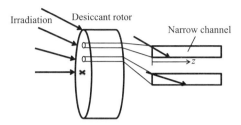

Fig. 2. 2　Image of direct heating of the desiccant rotor and incident angle

Previously, Hamamoto and Mori investigated direct solar heating of a rotor[1]. Their study aimed at making a theoretical model of performance analysis for a desiccant rotor of direct heating type driven by solar energy. In their model, a diffused reflection was assumed to estimate the absorbed energy distribution in the rotor channel, instead of regular reflection. However, in their theoretical model, no certified energy distribution information was shown, and they only provided some review results. Therefore, here, some experiments are performed to find the absorption rate distribution characteristics of absorbed irradiation energy in the narrow channels of desiccant rotor with direct heating. The measuring object is a narrow channel which is a part of desiccant rotor. In order to clarify the energy distribution carefully, the measurement is carried out under different conditions of the length of narrow channels, incident angle of irradiation, temperature and humidity.

## 2.1.2　Data reduction

The method of data reduction is as follows.

As shown in Fig. 2. 1, $I(0)$ denotes the detected energy before entering the channel, that is, the amount of irradiation energy in the front of the channel at $z = 0$. $I(z)$ expresses the detected energy after passing through the channel at the position $z = z$. Then, the irradiation energy absorption rate $g(z)$, that is, the ratio of the absorbed energy between $z = 0$ and $z = z$ to the inlet energy is given by the Eq. (2.1) for different lengths, incident angles and sample channels.

$$g(z) = 1 - \frac{I(z)}{I(0)} \qquad (2.1)$$

The energy absorption rate $g(z)$ is a cumulative value. For an infinitely long length of channel, the absorption rate $g(\infty)$ is equal to 1, and it means that all of the irradiation energy is absorbed by the channel.

In this study, the energy absorption rate $g(z)$ represents the irradiating energy distribution onto the wall of the narrow channel. The measuring error range of $I(0)$ and $I(z)$ is $\pm 5\%$ of measurement values, therefore, the error range of $g(z)$ has $\pm 10\%$.

### 2.1.3  Materials

Silica gel is widely usedas the desiccant material for dehumidification of air nowadays. Therefore, the present key sample channel were made by paper sheet containing silica gel. The sample channels, which corresponds to the part of the silica gel rotors, have a cross section of corrugate shape. Two shapes of the sample channels are used in this study; shape M and shape S as shown in Fig. 2.3. They rotate $\pi/2$ each other. The desiccant rotor is generally in the circular shape and its cross section is formed by a concentric circular multilayer with corrugated sheets. The irradiation ray from several directions comes to the rotor channels cross section. Therefore, both of the shape M and shape S are tested here.

This study also aims to find the influence factor on energy absorption in the narrow channels. So it is essential to select different materials for comparison with silica gel. As another sample, polypropylene plastic hollow board (hereinafter referred to as PP) is choiced as shown in Fig. 2.4. Compared to the silica gel channels wall, PP has smooth surface, and the channels shape of cross section is rectangular, simpler than that of silica gel channel. Because of a non−circular shape of the sample channels, a hydraulic diameter $D_e$ by Eq. (2.2) is introduced as an effective diameter for following data calculation.

$$D_e = \frac{4A}{S} \qquad (2.2)$$

where, $A$ is the cross−sectional area of single channel, and $S$ is the cross−sectional perimeter of single channel. These sizes are listed in the Tab. 2.1 described later.

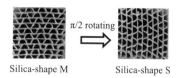

$\pi/2$ rotating

Silica-shape M          Silica-shape S

Fig. 2. 3    Different shapes of silica gel rotor element

Fig. 2. 4    Sample channels made by PP

## 2.1.4　Apparatus

A schematic diagram of direct heating of thesample channel is shown in Fig. 2. 5. The measurement apparatus is composed of a red laser, a goniometer, sample channels and a sensor of laser power meter. Energy incident angle $\theta$ and sample channel length $z$ are set as variables.

In this measurement, a single wavelength light is used as a source of irradiation instead of solar energy. In consideration of multi−wavelength of solar energy, a red laser is selected because its intensity of spectrum is relatively higher than others[2]. The wavelength of red laser is $6.5 \times 10^{-7}$ m.

To measure energy unabsorbed by the sample channel wall, the laser power meter is used as an energy receptor. The laser power meter is a convenience and usual approach for red laser measurement. Its maximum measuring error is ±5%. The diameter of sensor window is 0. 014m less than the width and height of desiccant samples (0. 016 m × 0. 020 m). The sensor window is covered by the desiccant sample.

To gain a better understanding of the distribution characteristics of absorbed energy in the sample narrow channels for different incident irradiation angles, the irradiation angle is changed in the range of $-4\pi/9$ to $4\pi/9$ and measured by the goniometer. The goniometer is set at the entrance position of the sample channel, and the laser is set in front of the sample channel. The distance between the laser and sample is set at 0. 15m.

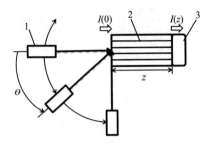

Fig. 2. 5　Schematic of the measuring apparatus for irradiation energy absorption rate in a sample channel

1—Laser　2—Sample　3—Laser power meter

## 2.1.5　Conditions

Tab. 2. 1 shows sample parameters and measurement ambient conditions. The measurement was performed in the temperature-humidity controlled room where the temperature and humidity of air can be controlled independently. The controlled temperature range was from 10℃ ± 1℃ to 50℃ ± 1℃ , and the humidity range was from 40% ±5% to 70% ±5%. The resistance thermometer was used to measure the room temperature. The humidity was measured by the sensor.

**Tab 2.1　Sample parameters and ambient conditions**

| Energy Source | Laser ( red ) | |
|---|---|---|
| Sample material | Silica gel( shape M , shape S )　$q^* =0.85\varphi/(1+4.5\varphi)$ | PP |
| Channel length $z$ /m | 0.005 , 0.010 , 0.015 | 0.005 , 0.010 ,　0.015 ,0.020 , 0.030 |
| Cross-sectional area of a channel $A$ /m$^2$ | $2.7\times10^{-6}$ | $4\times10^{-6}$ |
| Cross-sectional circumference $S$ /m | $7.7\times10^{-3}$ | $8\times10^{-3}$ |
| Channel hydraulic diameter $D_e$/m | $1.4\times10^{-3}$ | $2.0\times10^{-3}$ |
| Wall thickness /mm | 0.2 | 0.1 |
| Thirteen points height roughness /μm | 6.79 | 1.75 |
| Incident angle $\theta$ /rad | From $-4\pi/9$ to 0 and from 0 to $4\pi/9$ | |
| Temperature $T$ /℃ | 25 , 35 | |
| Relative humidity $RH$ /% | 40 , 60 | |

The temperature of the laser emitter was measured by an Infra-red thermometer. Usually an emitting power depends on the temperature of the laser emitter, but, here, the temperature was controlled to be almost constant during each measurement by keeping small emitting power and small measuring time. The rotor sample temperature was supposed almost equal to that of the controlled room air.

The equilibrium adsorption $q^*$ of a desiccant rotor material is an important parameter of the rotor. The equilibrium adsorption becomes larger at higher humidity. The equilibrium adsorption of the present rotor silica gel was already obtained in the measurement of the previous study[3] and listed in Tab. 2. 1.

From the preliminary test[4] , air in the channel has a tiny absorbing capacity to radiation energy. The goal of sample test is only to clarify the distribution characteristic of absorbed radiation energy in the narrow channels made by different materials, silica gel and PP.

## 2.1.6   Results and discussion

Measurement results of the energy absorption rate $g(z)$ for silica gel shape M and PP samples at temperature of $25°C$ and relative humidity of $40\%$ are shown in Fig. 2. 6, plotted against the incident angle $\theta$. It can be seen that the energy absorption rate shows a symmetrical distribution about the line of incident angle $\theta = 0$, no matter different channel materials, conditions and sample lengths. From this symmetrical result, it is enough to choose one side angle range from 0 to $4\pi/9$ in the following study.

When the incident angle is close to $\pi/2$, almost no incident ray can reach the inner wall of the sample channels. Hence, the absorption rate of irradiation energy is almost 0. For the same reason, at the irradiation angle $4\pi/9$, the incident ray reaching the channel inner wall becomes small, and the absorption rate of irradiation energy obviously reduces.

No energy should be absorbed when the incident angle is 0, because the irradiated energy will pass the channels from inlet to outlet directly. However, in fact, small portion of the energy is absorbed at the front edge of channel wall. Therefore, the energy absorption rate at angle of 0 is not 0, but smaller than that at the other incident angle.

In fact, the laser irradiation is not a single straight line, and has a straggling angle of $\pm\pi/2000$ rad. This means that the incidental angle $\theta$ of the laser has an error range of $\pm\pi/2000$ rad. Therefore, some irradiation is absorbed by the channel wall even at $\theta=0$.

Additionally, as the incident angle increases, the incident heat flux to the wall increases, although the direct irradiated area on the channel inner wall becomes small. Consequently, the entrance edge of the sample channel is expected to be heated strongly with increasing the irradiation incident angle.

As seen in Fig. 2.6, the absorption rate increases as the incident angle increases from 0, and is maintained at almost constant in the range from middle to high angle. The range of constant absorption rate becomes wide as the sample length increases, and occupies from $\pi/6$ to $7\pi/18$ at the length over 0.010 m for silica gel and 0.015 m for PP.

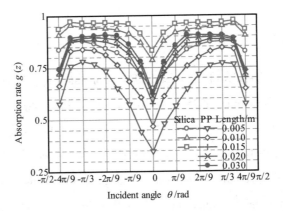

Fig. 2.6　Energy absorption rate comparison between symmetrical incident angles

The relation between the energy absorption rate $g(z)$ and the channel length is shown in Fig. 2.7. In the figure, four angles 0, $\pi/18$, $\pi/3$ and $4\pi/9$ are chosen as a parameter of the incident angle, and $\pi/3$ data represents the trend of energy absorption rate from the angles of $\pi/6$ to $7\pi/18$. It is revealed that the rate $g(z)$ generally increases with an increase of the length, and becomes constant at long length. In addition, the effect of temperature on the energy absorption rate is small both for the silica gel channel and PP channel. The results for different humidity conditions are shown in Fig. 2.8. The rate also doesn't depend on the humidity. Therefore, the ambient temperature and

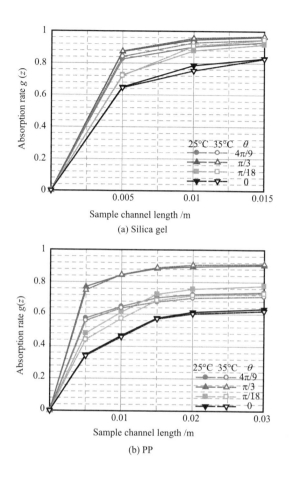

(a) Silica gel

(b) PP

Fig. 2. 7    Absorption rate comparison between different temperatures

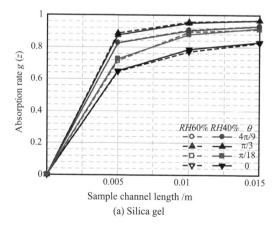

(a) Silica gel

Fig. 2. 8

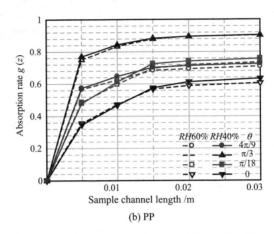

(b) PP

Fig. 2. 8    Absorption rate comparison between different humidity conditions

humidity have no influence on the energy absorption rate distribution. This means that there is no influence of the amount of adsorbed water on the energy absorption rate, and it is similar to the trend of the energy emission from the desiccant rotor material[5].

Fig. 2. 9 (a) shows the comparison of the energy absorption rate between the silica gel and PP sample. The absorption rate of silica gel is larger than that of PP at the same incident angle. There is diffused reflection in inner walls of a channel[1], and the diffused reflection depends on surface roughness. The roughness of silica gel channel wall is larger than that of PP. Hence, the diffused reflection occurs strongly in the silica gel channel compared with the PP channel. Therefore, the energy absorption rate of silica gel channel is larger than that of PP as shown in Fig. 2. 9 (a). Also, as shown in Fig. 2. 9 (b), there is almost no difference between shapes M and S of silica gel channel. Consequently, there is no need to consider any other shapes.

In addition, there is recognized other marked point in Fig. 2. 9 as to the channels length. When the channel length is more than 0. 015m, the energy absorption rate almost never changes, namely constant, and it means the amount absorbed is little in the range over 0. 015m length. Therefore, in conclusion, the channel length of 0. 015m is enough for considering the radiation energy absorption rate.

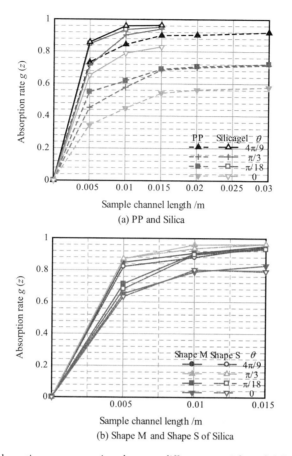

Fig. 2.9    Absorption rate comparison between different materials and different shapes

## 2.2    Energy distribution measurement of solar irradiation

### 2.2.1    Concept of measurement

In the measurement in the previous section, the red laser with the wavelength of 0.65 μm was used as an irradiation source. However, solar energy contains irradiations with multi-wavelengths, and absorbed energy might be greatly influenced by the wavelength. Therefore, it is better to clarify the distribution of absorbed solar irradiation energy in the narrow channels of desiccant rotor.

## 2.2.2 Data reduction

The method of data reduction is the same as that mentioned in the section 2.1.2.

## 2.2.3 Materials

The materials are also the same as those mentioned in the section 2.1.3.

## 2.2.4 Apparatus

The measurement apparatus is shown in Fig. 2.10, and composed of solar irradiation, basal plate, sample channels, laser power meter, nail, ruler, rotating holder, and tracking device (Nano. Tracker equatorial). Irradiation incident angle $\theta$ and sample channel length $L$ are set as variables. Fig. 2.11 shows photos of the apparatus for meas-

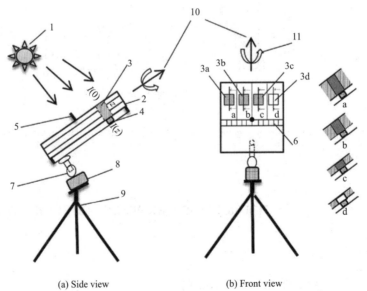

(a) Side view　　　　　　　　　　(b) Front view

Fig. 2.10　Schematics of measuring apparatus for solar irradiation energy absorption rate

1—Solar irradiation　2—Basal plate　3—Samples of different lengths (3a: $L = 0.015$m, 3b: $L = 0.01$m, 3c: $L = 0.005$m and 3d: no sample)　4— Laser power meter　5—Nail　6—Ruler　7—Rotating holder　8—Tracking system(Nano. tracker equatorial)　9—Tripod　10—North Stardirection　11—Solar tracking rotation

uring solar energy distribution.

To consider the influence of solar irradiation of different incident angles, the basal plate and samples can be rotated to different angles simultaneously by the manual controlling of incidental solar rays. The rotated angle is arranged by the shuttle length of nail as shown in Fig. 2.10, and changed in the range of 0 to $4\pi/9$. The starting incident angle is fixed so that the solar ray is vertical to the cross section of the sample. In this study, the tracking device (Nano. Tracker equatorial) was used to decrease the error of measurement due to the earth rotation.

Fig. 2.11    Photos of apparatus for measuring solar energy absorption rate

## 2.2.5   Conditions

Tab. 2.2 shows sample parameters and measuring conditions

**Tab. 2.2   Sample parameters and measurement conditions**

| Energy Source | Solar Irradiation | |
| --- | --- | --- |
| Sample Material | Silica gel $q^* = 0.85\varphi/(1 + 4.5\varphi)$ | PP |
| Channel length $z$ /m | 0.005, 0.010, 0.015 | |
| Cross-sectional area of a channel $A$ /m$^2$ | $2.7\times10^{-6}$ | $4\times10^{-6}$ |
| Cross-sectional circumference $S$/m | $7.7\times10^{-3}$ | $8\times10^{-3}$ |

( continued )

| Energy Source | Solar Irradiation | |
| --- | --- | --- |
| Channel hydraulic diameter $D_e$/m | $1.4 \times 10^{-3}$ | $2.0 \times 10^{-3}$ |
| Wall thickness /mm | 0.2 | 0.1 |
| Thirteen points height roughness /$\mu$m | 6.79 | 1.75 |
| Incident angle $\theta$ /rad | 0 to $4\pi/9$ | |
| Temperature $T$ /℃ | 7, 12.1 | |
| Relative humidity $RH$ /% | 42, 72 | |

## 2.2.6 Results and discussion

Measurements of the energy absorption rate $g(z)$ of solar irradiation for the silica gel and PP at temperature 12.1℃ and relative humidity 42% are shown by solid symbols in Fig. 2.12, plotted against the incident angle $\theta$. Like the case of the laser irradiation, the absorption rate is not zero at $\theta = 0$, and obviously reduces at $\theta = 4\pi/9$ close to $\pi/2$. Also, for silica gel, the absorption rate in the range of $\theta = \pi/9$ to $\theta = 7\pi/18$ shows almost constant for the channel length of 0.010m and 0.015m, while it gradually increases for the short length 0.005m. Compared to silica gel, the constant range of the absorption rate of PP is limited to the region near $\pi/3$, and generally the PP adsorption rate increases with the incident angles in the range of 0 to $\pi/3$.

In Fig. 2.13, the absorption rates are compared at different ambient humidity and temperature conditions. The effects of humidity and temperature of surrounding air on $g(z)$ are seen little even for the silica gel channels within the measurement accuracy. This result is the same as that of the laser light measurement, and it means that there is no influence of the amount of adsorbed water on the absorption rate.

Comparing between silica gel and PP, the energy absorption rate of silica gel is larger than that of PP as shown in Fig. 2.13, similarly to the case of laser irradiation measurement. This difference is supposed to be caused by the high diffused reflection in the silica gel narrow channels due to large surface roughness.

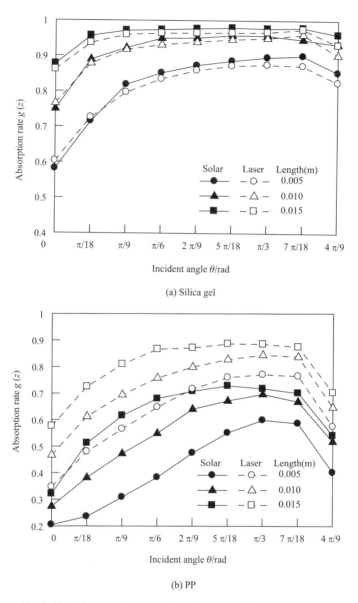

(a) Silica gel

(b) PP

Fig. 2. 12    Changes of energy absorption rate with incident angles

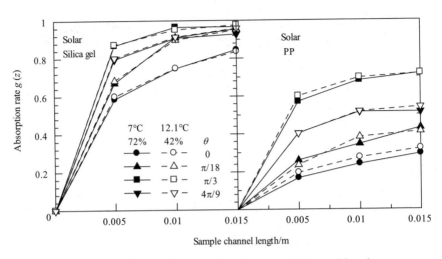

Fig. 2. 13 Changes of energy absorption rate with channel lengths

## 2.3 Comparison between laser and solar irradiations

For the comparison to the results for solar irradiation, the absorption rates for the laser light are depicted in Fig. 2. 12.

For the silica gel channels shown in Fig. 2. 12 (a), the absorption rate trend of solar irradiation is similar to that of the laser light. So, the wavelength of solar rays weakly influences the absorption rate of the silica gel desiccant rotor.

However, for the PP channels shown in Fig. 2. 12 (b), the absorption rate of the laser is larger than that of the solar. To get behind this difference between solar irradiation and laser light, the transmissivity of PP material was measured with UV–VIS spectrophotometer. Fig. 2. 14 shows its results. The transmissivity of PP increases with increasing the wavelength. The result indicates that a long wavelength ray in solar irradiation is less absorbed to the channel wall. According to the evaluation considering the spectral energy change which depends on the wavelength, the average transmissivity of PP is larger for solar rays than for red laser. Therefore, the absorption rate of solar irradiation in the PP channel is lower than that of laser light.

36

The transmissivity of silica gel is very small, only 0.4%[6]. Therefore, the absorption rate in the silica gel channel is maintained large for solar irradiation.

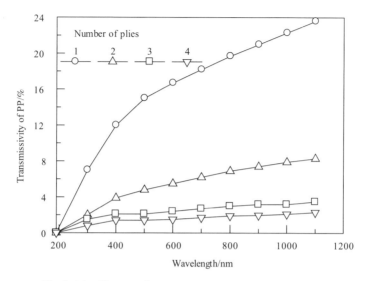

Fig. 2.14　Changes of transmissivity of PP with wavelengths

## 2.4　Conclusions

In this chapter, the measurement and clarification of the absorption rate distribution characteristics of absorbed irradiation energy in the narrow channel wall of desiccant rotor with direct heating were conducted. As an energy source, laser light and solar irradiations were selected by different irradiation devices. Two kinds of materials, silica gel and PP, were used. The measurement was carried out under the different conditions. The main findings are as follows.

(1) The absorption rate of irradiation energy changes with the irradiation incident angles, leading to the energy distribution depending on the incident angle.

(2) Both for laser and solar irradiations, the energy absorption rates of silica gel and PP do not depend on the temperature and humidity, namely, there was no influence of the amount of adsorbed water on the absorption rate. The energy absorption rate of

silica gel was larger than that of PP. This difference is supposed to result from a strong diffused reflection on the channel walls due to large surface roughness of silica gel.

(3) For silica gel, the absorption rate trend of solar irradiation was similar to that of laser light. Therefore, the influence of the solar ray wavelength on the energy absorption rate of silica gel is weak. For PP material, however, the absorption rate of solar irradiation was lower than that of laser light. This is because the long wavelength ray in solar irradiation was less absorbed to the wall due to high transmissivity of PP material.

# References

[1] HAMAMOTO Y, MORI H. A model of absorbed energy distribution and numerical simulation in a desiccant rotor regenerated by concentrated solar irradiance [J]. Proceedings of Renewable Energy 2010, 33,1-4.

[2] INCROPERA F P, DEWITT D P, BERGMAN T L, et al. Fundamental of heat and mass transfer sixth edition[C]. John Wiley & Sons, 2005, 724-726.

[3] KASHIMA T. Study on the adsorption/desorption characteristics of water vapour adsorbent in air flow channel wall regenerated by concentrated solar irradiation [D]. Fukuoka, Kyushu University, 2015.

[4] LI J, HAMAMOTO Y, MORI H. Measurement and prediction of absorbed irradiation energy distribution in narrow channel of desiccant rotor[C]. Proceedings of 15th International Heat Transfer Conference, 2014.

[5] HAMAMOTO Y, MORI H, AKAI T, et al. Measurement of emissivity of adsorbent material for development of an adsorption humidity control system with regeneration process by concentrated irradiation[C]. Proceedings of the International Sorption Heat Pump Conference 2008, AB-058.

[6] HAMAMOTO Y, MORI H, AKAI T, et al. Measurement of a spectral reflectivity, absorptivity and transmissivity of water vapour adsorbents[C]. Proceedings of the 46th National Heat Transfer Symposium of Japan, 2009,G-1103, 217-218.

# Chapter 3　Predictions for energy distribution

The characteristics of irradiation energy absorption rate distribution were shown for different materials narrow channels in Chapter 2. In this chapter, three models are introduced to predict the energy absorption rate in a narrow channel, and the prediction accuracy is examined compared with the measurements in Chapter 2.

## 3.1　Concept of prediction models

### 3.1.1　Prediction model regarding the energy absorption efficiency

In the model of previously developed in document[1], called "RE2010" model here, the concentrated solar ray lights on the rotor element, and continues the reflection and diffusion in the narrow channel with the energy absorption on the element. The concept of the energy absorption process was described in the Chapter 1. In the model, the energy absorption efficiency $f_a(z)$ is given in Eq. (3.1)

$$
\begin{aligned}
z \leqslant L_{di} &: f_a(z) = a + \phi(1-a) \\
z > L_{di} &: f_a(z) = \{a + \phi(1-a)\}(1-\phi)^{Int(z/L_{di})}(1-a)^{Int(z/L_{di})} = E\exp(-Bz)
\end{aligned}
\tag{3.1}
$$

where $a$ is the absorptivity of desiccant rotor and $\phi$ is the diffused reflection ratio, and $a$ and $\phi$ are assumed 0.2 and 0.9, respectively[1]. The distance in the flow direction is denoted by $z$, and the direct heating distance in the inlet portion is by $L_{di}$. The power part $Int(z/L_{di})$ represents the number of reflection in the channel, and $B$ and $E$ are both a function of these parameters.

The validity confirmation of this model is shown in Fig. 3.1. For comparison, experiment data of the energy absorption rate in the laser measurement in Chapter 2 are used. The material is silica gel. The $r/R$ in Fig. 3. 1 means a non-dimensional radius of the desiccant rotor corresponding to the irradiation angle $\theta$ when the lens size, rotor diameter and distance between them are fixed. The results of the "RE2010" model and the present experiment data both show the decrease trend of absorption efficiency $f_a(z)$ with the rotor thickness, like a exponential function. Therefore, the model is tenable to some extent, and deserves to be used as a prediction model for evaluating the energy absorption rate in this study.

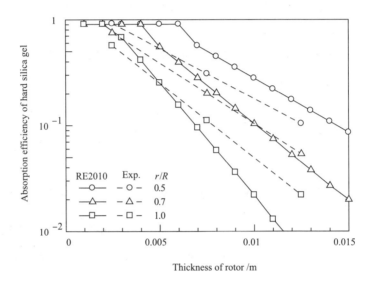

Fig. 3. 1   Comparison of "RE2010" model predictions with experimental data

As shown in Fig. 3. 1, for $z > L_{di}$, the energy absorption efficiency can be reproduced by a simple formula like $E \exp(-Bz)$. The form of the equation is discussed as follows.

Fig. 3. 2 shows an irradiation energy absorption in a narrow channel. The amount of irradiation energy absorbed between the interval $\Delta z$ is given as the left-hand side of Eq. (3.2). The absorption energy amount is also written using the absorption efficiency $f_b(z)$ as the right-hand side of Eq. (3.2).

$$[I(z - \Delta z) - I(z)] \frac{\pi}{4} D_e^{\,2} = I(0) f_b(z) \Delta z \pi D_e \qquad (3.2)$$

Combining with Eq. (2.1) of the energy absorption rate $g(z)$ in Chapter 2, $f_b(z)$ is rewritten as Eq. (3.3).

40

$$f_b(z) = \frac{D_e}{4} \frac{g(z) - g(z - \Delta z)}{\Delta z} \tag{3.3}$$

The relationship between $f_a(z)$ for flat plane path and $f_b(z)$ for the cylindrical narrow channel is expressed as Eq. (3.4).

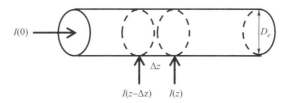

$I(0)$ $\qquad$ $D_e$

$I(z{-}\Delta z)$ $\quad$ $I(z)$

$\Delta z$

Fig. 3.2    Energy absorption in the interval of $\Delta z$

$$f_a(z) = \frac{4}{D_e} f_b(z) L_{di} = \left[ \frac{g(z) - g(z - \Delta z)}{\Delta z} \right] L_{di} \tag{3.4}$$

According to the equation structure of Eq. (3.1), the equation form of Eq. (3.5) is suitable for the energy absorption efficiency $f_a(z)$ when $z > L_{di}$

$$f_a(z) = E \exp(-Bz) \tag{3.5}$$

Theenergy absorption efficiency $f_a(z)$ can be calculated from the measurement data. Thence, the parameters $B$ and $E$ are determined by the measurement data. The energy absorption efficiency $f_a(z)$, Eq. (3.1), will be used in a numerical simulation for evaluating the performance of the desiccant rotor in Chapter 5.

### 3.1.2    Prediction model of view factor

In a radiation exchange, the view factor $F_{ij}$ is defined as the fraction of the radiation leaving surface $i$ intercepted by surface $j$. From the definition of the view factor, the summation rule is defined as Eq. (3.6), when applied to each of the $N$ surfaces in the enclosure. The term $F_{ii}$ called self-view factor is equal to $0$[2].

$$\sum_{j=1}^{N} F_{ij} = 1 \tag{3.6}$$

In the model of view factor, the process of radiation energy absorbing in a narrow channel is treated as the same character as the view factor. The energy absorption process in the narrow channel is shown in Fig. 3.3 (a) with different incident angles,

and the view factor is shown in Fig. 3.3 (b). The view factor $F_{12}$ is similar to the unabsorbed energy ratio at exit end of the narrow channel, namely, $F_{12}$ corresponds to the ratio $I(z)/I(0)$. Thus, the view factor $F_{13}$ ($= 1 - F_{12}$) corresponds to the energy
absorption rate $g(z)$. It is tried to express the energy absorption rate $g(z)$ using the
model of view factor.

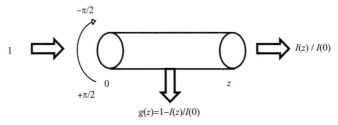

(a) Energy absorption of narrow channel

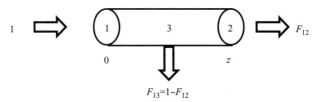

(b) Energy absorption of view factor

Fig. 3.3   Comparison between energy absorption process and view factor

The prediction method for the model of view factor is as follows.

The absorbed energy in the length $\Delta z$ is expressed in the left-hand side of
Eq. (3.7) where $S$ is the cross-sectional perimeter of the channel. The absorbed
energy is related to the difference of the view factor in the distance $\Delta z$ as the right-hand
side of Eq. (3.7).

$$I(0) f_b(z) \Delta z S \propto F_{13}(z) - F_{13}(z - \Delta z) \tag{3.7}$$

Hence, the energy absorption efficiency $f_b(z)$ is written as Eq. (3.8), approxi-
mately to $dF_{13}/dz$. Eq. (3.10) is derived from Eq. (3.9).

$$f_b(z) \propto \frac{F_{13}(z) - F_{13}(z - \Delta z)}{\Delta z} \approx \frac{dF_{13}}{dz} \tag{3.8}$$

$$f_b(z) \propto \frac{g(z) - g(z - \Delta z)}{\Delta z} \tag{3.9}$$

$$\frac{dF_{13}}{dz} \propto \frac{g(z) - g(z - \Delta z)}{\Delta z} \tag{3.10}$$

Considering the difference of the absorption rate $g(z) - g(z - \Delta z)$ with view factor $dF_{13}/dz$, fitting parameters $\beta(z)$ and $x(z) = \beta(z)/\Delta z$ are introduced as Eq. (3.11) and Eq. (3.12) as a function of $z$.

$$\xi(z) \frac{dF_{13}}{dz} = \frac{g(z) - g(z - \Delta z)}{\Delta z} \tag{3.11}$$

$$\beta(z) \frac{dF_{13}}{dz} = g(z) - g(z - \Delta z) \tag{3.12}$$

$$f_b(z) = \frac{D_e}{4} \frac{g(z) - g(z - \Delta z)}{\Delta z} = \frac{D_e}{4} \frac{\beta(z)}{\Delta z} \frac{dF_{13}}{dz} = \frac{D_e}{4} \xi(z) \frac{dF_{13}}{dz} \tag{3.13}$$

The differentiation $dF_{13}/dz$ in the right-hand side of Eq. (3.13) is given by Eq. (3.14)[2].

$$\frac{dF_{13}}{dz} = -\frac{4}{D_e^2} z + \frac{2}{D_e} \left[ \left( 1 + \frac{z^2}{D_e^2} \right)^{\frac{1}{2}} + \frac{z^2}{D_e^2} \left( 1 + \frac{z^2}{D_e^2} \right)^{-\frac{1}{2}} \right] \tag{3.14}$$

According to the rule of view factor $F_{13} = 1 - F_{12}$, $dF_{13}/dz$ is equal to $-dF_{12}/dz$. The view factor $F_{12}$ for the cylindrical channel is given by the following Eq. (3.15)[2].

$$F_{12} = \frac{1}{2} \left[ X - \sqrt{X^2 - 4\left( \frac{R_2}{R_1} \right)^2} \right] \tag{3.15}$$

where, $R_1 = \frac{r_1}{z}$, $R_2 = \frac{r_2}{z}$, $X = 1 + \frac{1 + R_2^2}{R_1^2}$ and $r_1 = r_2 = D_e/2$.

The energy absorption efficiency and fitting parameters $\beta(z)$ and $\zeta(z)$ are determined from Eq. (3.14) based on the measurements.

### 3.1.3　Prediction model of Beer's law

Spectral radiation absorption in a gas (or in asemi-transparent liquid or solid) is a function of the absorption coefficient $\kappa_\lambda(1/m)$ and the thickness $L(m)$ of the gas medium[3]. Beer's law is a useful tool in approximate radiation analysis. Then, the absorptivity $\alpha_\lambda$ is expressed as Eq. (3.16).

$$\alpha_\lambda = 1 - e^{-\kappa_\lambda L} \tag{3.16}$$

In Beer's law, $\alpha_\lambda$ denotes the absorptivity of gas layer and $\kappa_\lambda$ is the absorption coefficient of gas layer. The process of energy absorption in a channel is similar to that of radiation absorption in a gas layer. So, the prediction equation of $g(z)$ can be expressed in the similar form to Eq. (3.16) as Eq. (3.17).

$$g(z) = 1 - e^{-\kappa z} \qquad (3.17)$$

In Beer's law, the absorption coefficient $\kappa_\lambda$ of gas layer is constant. Similarly, the coefficient $\kappa_\lambda$ in Eq. (3.16) is expected to be constant. The coefficient $\kappa_\lambda$ will be checked for different materials channels using the experimental data.

## 3.2 Parameters for silica gel sample and accuracy evaluation

### 3.2.1 "RE2010" model

In the predicted equation $f_a(z)_{pred}$ of the "RE2010" model, two variable parameters $B$ and $E$ are introduced as in Eq. (3.1). The parameters $B$ and $E$ are affected by the incident angle of radiation energy. It is necessary to determine the accurate values of these parameters based on the experimental data in Chapter 2. As shown in Fig. 2.12(a), the energy absorption rate of silica gel is almost the same between solar irradiation and laserlight. Therefore, the fitting parameters are determined using the laser measurement data, and they can reproduce the absorption rate of solar experiment well. In this way, as a function of the incident angle $\theta$, the fitting parameters of $B$ and $E$ were determined for silica gel material using the experimental data shown in Fig. 2.6. The determined functions are shown in Tab. 3.1.

Tab. 3.1　Fitting parameters $B$ and $E$

| | $\pi/36 \leqslant \theta < \pi/6$ | $\pi/6 \leqslant \theta \leqslant 4\pi/9$ |
|---|---|---|
| $B$ | $573\theta + 100$ | $400$ |
| $E$ | $-1.43\theta + 1.12$ | $1.55e^{-2.55\theta}$ |

Fig. 3.4 shows the comparison between the model prediction $f_a(z)_{pred}$ and the

experimental data $f_a(z)_{exp}$. In the figure, experimental data of energy absorption efficiency $f_a(z)_{exp}$ was calculated values from measured data of the adsorption rate $g(z)_{exp}$ using Eq. (3.4). The prediction reproduces all the $f_a(z)_{exp}$ data both for solar and laser irradiations well.

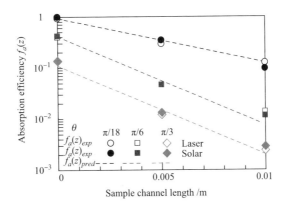

Fig. 3.4    Comparison of $f_a(z)$ between experiments and predictions of silica gel material

The energy absorption efficiency $f_a(z)_{pred}$ will be used for the prediction of desiccant rotor dehumidification performance in Chapter 5.

To examine the accuracy of the "RE2010" model, the predicted absorption rate $g(z)_{pred}$ by Eq. (3.18) was compared with the measured $g(z)_{exp}$.

$$g(z)_{pred} = \frac{f_a(z)_{pred}}{L_{di}}\Delta z + g(z - \Delta z)_{pred} \tag{3.18}$$

Here, $\Delta z = 0.005$ m from the measurement, and $g(0)_{pred} = 0$.

Fig. 3.5 shows the comparison of $g(z)_{pred}$ and $g(z)_{exp}$ for the laser measurement. The maximum deviation between $g(z)_{exp}$ and $g(z)_{pred}$ is almost equal to the measurement error $\pm 10\%$. In addition, the comparison for the solar measurement shows the same result, as shown in Fig. 3.6. Consequently, it was confirmed this prediction method was feasible in practice.

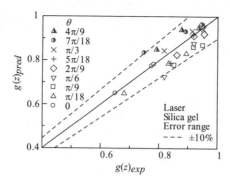

Fig. 3.5  Comparison of $g(z)$ between experiments and predictions of silica gel material with

aser measurement (Laser is the energy power, Silica gel is the material of narrow channel)

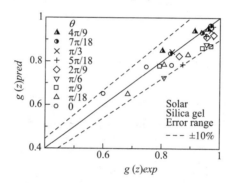

Fig. 3.6  Comparison of $g(z)$ between experiments and predictions of silica gel material with

solar measurement (Solar is the energy power, Silica gel is the material of narrow channel)

## 3.2.2  View factor model

For the view factor model, the fitting parameters $\beta(z)$ and $\zeta(z)$ are determined based on the data shown in Fig. 2.6, and expressed by Eq. (3.19) and Eq. (3.20) as a function of the channel length coordinate $z$, not influenced by the irradiation incident angle.

$$\beta(z)_{pred} = 1650z + 0.67 \tag{3.19}$$

$$\xi(z)_{pred} = \frac{\beta(z)_{pred}}{\Delta z} = 3.30 \times 10^5 z + 134 \tag{3.20}$$

Similarly to the case of the "RE2010" model, each fitting parameter is given by the same equation both for laser and solar measurements.

To examine the accuracy of the view factor model, the predicted absorption rate $g$ $(z)_{pred}$ calculated by Eq. (3.21) was compared with the measured $g(z)_{exp}$.

$$g(z)_{pred} = \beta(z)_{pred} \frac{\mathrm{d}F_{13}}{\mathrm{d}z} + g(z - \Delta z)_{pred} \qquad (3.21)$$

Fig. 3.7 and Fig. 3.8 show the comparison results between $g(z)_{exp}$ and $g(z)_{pred}$ for laser and solar measurements, respectively. They indicate that the deviation is less than or within the measurement error ± 10%. Consequently, it was confirmed the prediction model of view factor was feasible.

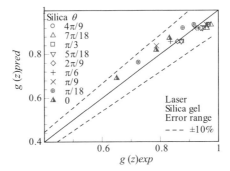

Fig. 3.7　Comparison of $g(z)$ between experiment and view factor model of silica gel material with laser measurement (Laser is the energy power, Silica gel is the material of narrow channel)

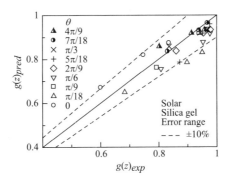

Fig. 3.8　Comparison of $g(z)$ between experiment and view factor model of silica gel material with solar measurement (Solar is the energy power, Silica gel is the material of narrow channel)

### 3.2.3 Beer's law model

For the Beer's law model, the values of $\kappa$ determined from the silica gel data in Fig. 2.6 are shown in Fig. 3.9. It increases from 0 to a certain value with increasing the length in the range of 0 to 0.005m, while after then it continuously decreases. Hence, the coefficient $\kappa$ is not constant whatever on the whole or segmentation. Therefore, it is thought that the prediction of the Beer's law model is not suitable to use.

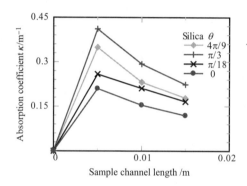

Fig. 3.9　Absorption coefficient $\kappa$ of silica gel material

## 3.3　Comparison of models

Among the three models, the prediction accuracies both of the "RE2010" model and view factor model are good and almost the same, while the Beer's law model is invalid. The prediction of the view factor model is simpler than the "RE2010" model. In the view factor model, it does not need to consider the influence of the irradiation angle. Finally, both of the "RE2010" model and view factor model are useful to predict the energy absorption rate. These two models will be used to simulate the performance of a desiccant rotor in Chapter 5.

The results for PP sample material are shown in the Appendix. The validity of the view factor model was confirmed, and two fitting parameter equations in the view factor model were provided for predicting the energy absorption rate of laser and solar irradiations.

## 3.4   Conclusions

Three models were presented for predicting the energy absorption rate: the "RE2010" model, view factor model and Beer's law model. The fitting parameters of each model were decided from the measurement results shown in Chapter 2. The following findings were also obtained.

(1) Both of the "RE2010" model and view factor model have almost the same good reproduction accuracy less than the maximum measurement error ±10%. Therefore, both models are feasible for the predicting energy absorption rate.

(2) The prediction method of the view factor model is simpler than the "RE2010" model due to no need to consider the irradiation incident angle.

(3) The model of Beer's law is invalid.

(4) For PP material, the validity of the view factor model was confirmed.

## References

[1]   HAMAMOTO Y ,MORI H. A model of absorbed energy distribution and numerical simulation in a desiccant rotor regenerated by concentrated solar irradiance [C]. Proceedings of Renewable Energy, 2010,33,1−4.

[2]   INCROPERA F P, DEWITT D P, BERGMAN T L, et al. Fundamental of heat and mass Transfer [C]. Fifth edition, USA: John Wiley & Sons, 2002, 816−822.

[3]   INCROPERA F P, DEWITT D P, BERGMAN T L, et al. Fundamental of heat and mass Transfer [C]. Sixth edition, USA: John Wiley & Sons, 2005, 842−847.

# Chapter 4 Experiment for measuring water vapour adsorption/desorption rate in desiccant rotor

In this chapter, experimental results of dehumidifying (adsorption) rate of a desiccant rotor both in adsorption and desorption processes are presented, and the influences of operating parameters on the dehumidifying rate and outlet humidity of the desiccant rotor are discussed. The measurement data will be used for comparison with numerical simulation in Chapter 5.

## 4.1 Measurement system

### 4.1.1 Apparatus

Fig. 4. 1 schematically shows the measurement apparatus composed of solar light, two Fresnellenses, test section, ducts for connecting each component, sensors for measuring temperature, humidity and air flow velocity, and air conditioning system (air generator). An opaque slide shutter set in front of the Fresnel lens isused to prevent the solar light irradiation onto the rotor during the adsorption process. When the shutter is closed, the light irradiation is blocked. Otherwise, when the shutter is unclosed, the rays from the light is concentrated through the Fresnellens to heat the rotor directly during the desorption process. Then, the desiccant rotor is regenerated.

Measurements are carried out under the different conditions of the inlet air flow velocity, temperature and humidity. The inlet temperature, relative humidity and velocity of air flow are measured at the duct A as shown in Fig. 4. 1. The outlet temperature and humidity of air flow are measured at the duct D. The outlet air humidity is an important

50

measuring indicator to evaluate the adsorption or desorption process in the desiccant rotor. Using the outlet air humidity, the dehumidifying rate of the rotor is calculated.

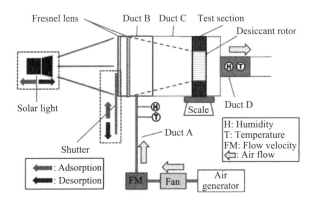

Fig. 4. 1    Schematic of measurement apparatus for dehumidifying rate of desiccant rotor

The Fresnel lens as shown in Fig. 4. 2 is used to concentrate the irradiation from the solar light and provide stronger heat flux to the desiccant rotor. Two lensesare used to change the direction of the concentrated solar light irradiation. Passing through the first Fresnellens, the diffused light raysare changed to the parallel rays. Then, after the second lens, the raysare concentrated on the rotor. The size of each Fresnellens is 0. 4m ×0. 4m and its thickness is about 3mm. The concentrated ratio is set at 5 from the distance between the lens and rotor front. The power of solar light is 500W, and part of the light rays reach the lens. The amount of the energy getting to the lens was estimated 52. 8W.

Usually, for a cyclical change of adsorption and desorption processes, the desiccant rotor rotates continually. Hence, the rotor material is cyclically exposed to the air of a different condition corresponding to the adsorption or desorption mode. In the present experiment, the rotor is not rotated, but the cycle adsorption/desorption processes is achieved by the periodic shutter opening and closing operation. This operation can simulate the rotor situation irradiated by the solar rays intermittently, namely, only during the desorption process.

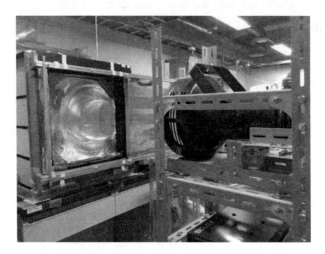

Fig. 4. 2   Solar light and Fresnellens in the apparatus

## 4. 1. 2　Material

Fig. 4. 3 shows the desiccant rotor made of silica gel material used in Chapter 2. The cross section of the rotor is formed from concentric circular–multilayer corrugated sheets shown in Fig. 4. 4.

Fig. 4. 3   Desiccant rotor in the measurement apparatus

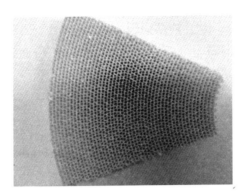

Fig. 4. 4    A part of silica gel desiccant rotor

## 4. 1. 3   Conditions and data evaluation method

Experimental conditions such as the rotor parameters and inlet air conditions are listed in Tab. 4. 1.

Tab. 4. 1   Sample parameters and experimental conditions

| Sample Material | Silica Gel Rotor<br>Equilibrium adsorption $q^* = 0.85\varphi/(1 + 4.5\varphi)$ [1] |
|---|---|
| Rotor thickness $L$/m | 0. 017 |
| Rotor mesh height /m | 0. 002 |
| Rotordiameter $D$/m | 0. 18 |
| Temperature $T$/K | 293, 298, 303 |
| Relative humidity $RH$/% | 50, 55, 60 |
| Air flow velocity $u_a$/(m $\cdot$ s$^{-1}$) | 0. 09, 0. 15, 0. 22 |
| Concentrated ratio of Fresnel lens | 5 |
| Adsorption cycle time /s | 180 |
| Desorption cycle time /s | 180 |

In the measurement, the dehumidifying rate ( $\Delta M_{ad}$ ) of the desiccant rotor is obtained by Eq. (4.1), given per hour.

$$\Delta M_{ad} = \rho_a u_a A(x_{in} - x_{out}) \times 3600 \qquad (4.1)$$

where $\rho_a$ is the density of air flow ($kg/m^3$), $u_a$ is the air flow velocity in front of the rotor ($m/s$), $A$ is the cross section area in front of desiccant rotor ($m^2$), and $x_{in}$ and $x_{out}$ are the inlet and outlet humidity, respectively.

The rate is calculated using the averaged ($x_{in} - x_{out}$) during the adsorption process. The dehumidifying rate $\Delta M_{ad}$ is equal to the amount of adsorbed water vapour of desiccant rotor in the cycle.

The results are obtained in the last cycle after reaching a steady state cycle of adsorption and desorption processes. Measuring errors of the relative humidity is $RH$ $\pm 1\%$, air flow velocity is $\pm 0.1$ m/s, temperature is $\pm 0.2$ K, respectively. Hence, the estimated measuring error is $\pm 0.3$ g/kg(DA) for the humidity and about $\pm 30\%$ for the dehumidifying rate.

## 4.2　Measurements results and discussion

### 4.2.1　Temperature and humidity measurements

Measurement results of the time variation of the outlet air humidity and temperature are shown in Fig. 4.5 and Fig. 4.6, respectively. The humidity decreases from the time intervla 0 to 180s in the adsorption process, while it increases from the time intervla 180s to 360s in the desorption process. The trends both of adsorption process and desorption process are almost symmetry. At the beginning of the adsorption the humidity decreases markedly and after reaching a minimum value it slightly increases. While, at the beginning of the desorption it increases and it decrease slowly after reaching a maximum value. It is recognized that the areas between $x_{out}$ and $x_{in}$ during the adsorption and desorption processes are almost the same, that is, the adsorption rate is equal to desorption rate in the cycle. The outlet air temperature shows entirely the trend of decrease in adsorption process and increase in desorption process, though fluctuating within 1K.

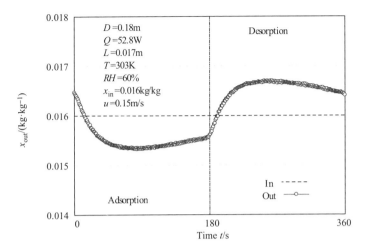

Fig. 4. 5   Outlet humidity both in adsorption and desorption process

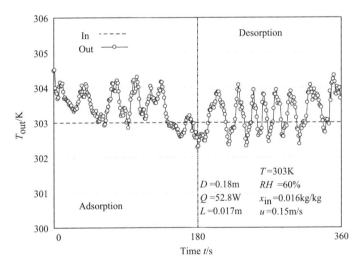

Fig. 4. 6   Outlet air temperature both in adsorption and desorption process

## 4.2.2  Influence of inlet air temperature, humidity and humidity ratio on the dehumidifying rate

The influence of the inlet air temperature on the dehumidifying rate $\Delta M_{ad}$ is shown in Fig. 4. 7, when the relative humidity is 60% and the air flow velocity is 0. 15m/s. The dehumidifying rate increases with increasing the inlet temperature. The higher temperature corresponds to the higher partial pressure of water vapour. Therefore, the equilibrium adsorption becomes larger at higher temperature. In addition, the increase rate from $T = 293$ K to $T = 298$ K is larger than that from $T = 298$ K to $T = 303$ K. This is because, when the desiccant is cooled adequately at low temperature, the adsorption rate becomes high, hence leading to a lower increasing rate at high temperature.

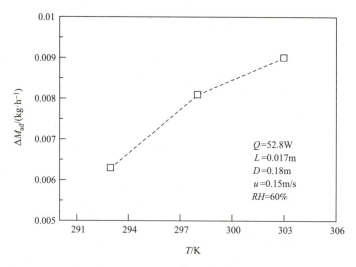

Fig. 4. 7　Influence of inlet air temperature on the dehumidifying rate of desiccant rotor

As shown in Fig. 4. 8, when the relative humidity of the inlet air increases, the dehumidifying rate increases lineally. This is because high humidity results in large potential for moisture transfer, that is, high partial pressure of water vapour, which provides more moisture removal capacity. Consequently, the desiccant rotor obtains better dehumidification capacity with higher humidity. The influences of inlet air temperature and humidity found here are similar to the results of other studies[2-3].

Fig. 4. 9 and Fig. 4. 10 show the influences of the inlet humidity $x_{in}$ and the inlet air enthalpy $h_{in}$ on the dehumidifying rate, respectively. The dehumidifying rate increases with increasing the humidity and enthalpy. These are also similar trends in the usual desiccant system.

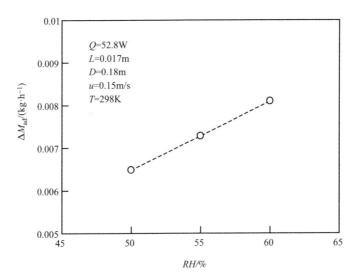

Fig. 4. 8   Influence of inlet air relative humidity on the dehumidifying rate of desiccant rotor

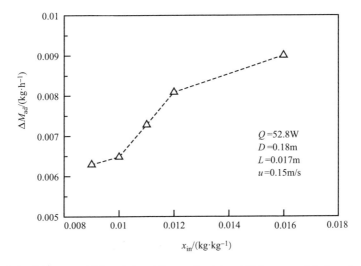

Fig. 4. 9   Influence of inlet air humidity on the dehumidifying rate of desiccant rotor

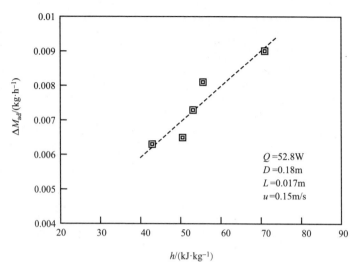

Fig. 4. 10   Influence of inlet air enthalpy on the dehumidifying rate of desiccant rotor

## 4.2.3   Influence of air flow velocity

Fig. 4. 11 shows the influence of air flow velocity on the time change of outlet humidity. The result shows that the outlet humidity variation becomes obviously small with increasing the air flow velocity from 0. 09m/s to 0. 22m/s.

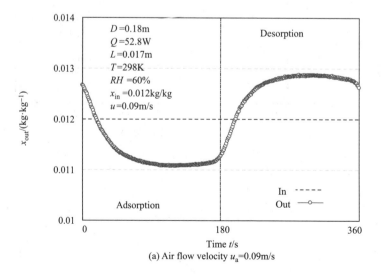

(a) Air flow velocity $u_a$=0.09m/s

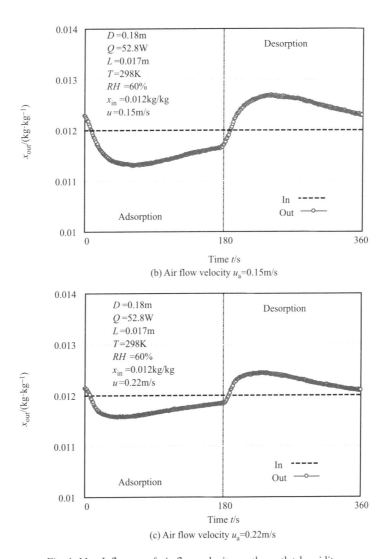

Fig. 4. 11    Influence of air flow velocity on the outlet humidity

The influence of the air flow velocity on the dehumidifying rate is shown in Fig. 4. 12. As expected from the time changes of the outlet humidity, the dehumidifying rate decreases with increasing the air flow velocity. As the air velocity increases, the residence time of the flowing air in the rotor channel is shortened. Therefore, from the result, it is supposed that, at high velocity, it is not enough time for the adsorbent to adsorb water vapour. On the other hand, it is expected that total amount of water vapour flowing in the channel decreases at significantly small air flow velocity. Hence, there is an optimum velocity to reach a maximum of the dehumidifying rate as seen in the usual desiccant system[4].

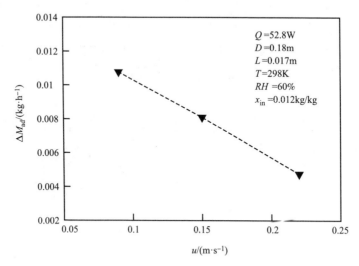

Fig. 4.12 Influence of air flow velocity on the dehumidifying rate of desiccant rotor

## 4.3 Conclusions

The measurements results of the dehumidifying rate in the desiccant rotor operating cyclically the adsorption and desorption processes are presented. The influences of inlet air temperature, relative humidity and flow velocity are examined. The main findings are as follows.

(1) The dehumidifying rate was affected by the inlet air temperature. It increased with increasing the temperature, because the high temperature gives high partial vapc pressure. In addition, the dehumidifying rate increased slowly in relatively high temperature condition. This is because, at low temperature, the desiccant is cooled adequately, hence leading to high adsorption capacity.

(2) As the relative humidity increased, the dehumidifying rate increased lineally. This is because partial pressure of water vapour became high.

60

(3) It was examined that the dehumidifying rate increased with the inlet humidity and enthalpy as with the usual desiccant system.

(4) The dehumidifying rate decreased with increasing the air flow velocity. This is because, at high velocity, there was no enough time for the adsorbent to adsorb water vapour.

# References

[1]　KASHIMA T. Study on the adsorption/desorption characteristics of water vapour adsorbent in air flow channel wall regenerated by concentrated solar irradiation [D]. Fukuoka. Kyushu University, 2015.

[2]　JIA C X, DAI Y J, WU J Y, et al. Experimental comparison of two honeycombed desiccant wheels fabricated with silica gel and composite desiccant material[J] Energy Conversion and Management, 2006, 47, 2523−2534.

[3]　WHITE S D. Characterization of desiccant wheels with alternative materials at low regeneration temperatures[J]. International Journal of Refrigeration, 2011, 34, 1786−1791.

[4]　HAMAMOTO Y, OKAJIMA J, MATSUOKA F , et al. Performance analysis of rotary dehumidifier/humidifier and systems, 1st report: Theoretical model[J]. Transactions of the Japan Society of Refrigerating and Air Conditioning Engineers, 2002, 19(3), 281−292.

# Chapter 5   Simulation of dehumidifying process of desiccant rotor

## 5.1   Mathematical analysis

### 5.1.1   Physical model

A cross-sectional view of the desiccant rotor is shown in Fig. 5.1. The corrugate or honeycomb type rotor consists of numerous narrow flow channel elements with a small equivalent diameter. The channel walls are made up of porous materials impregnated with adsorbent (desiccant material). In this study, silica gel was selected as the desiccant material due to its strong adsorbing ability and low regeneration temperature.

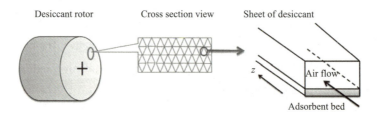

Desiccant rotor        Cross section view        Sheet of desiccant

Air flow

Adsorbent bed

Fig. 5.1   Rotor element and flat plate model

Both in adsorption process and desorption process, heat and mass are transferred simultaneously between air flow and channel wall in each narrow channel. For simplicity and easy understanding, the element is considered as a flat plate with a rectangular cross section and divided into two parts; air flow path and adsorbent bed as shown in the right diagram of Fig. 5.1. When humid air passes through the air flow path, some amount of water vapour can be adsorbed by the adsorbent, which realizes dehumidification.

The adsorbent adsorbing some amount of water vapour in the desorption process needs to be heated to a high temperature level to regenerate. The schematic of a direct heating type desiccant rotor using Fresnel lens is shown in Chapter 1. The solar irradiation concentrated with Fresnel lens impinges on the front surface of the desiccant rotor. Then, although a small part of the solar irradiation is absorbed by the front surface wall directly, other major part enters the narrow channel. In the channel, diffused reflection continuously occurs because the channel wall surface is not smooth. All the solar rays absorbed in the rotor channel are used to increase the channel wall temperature and to regenerate the desiccant material in the desorption process. The remaining rays pass through the channel and flow outside the rotor. Therefore, the path in the axis $z$ of the air flow direction should be divided into many small path intervals in the calculation. Heat and mass transfer between the air flow and adsorbent wall are considered in each small path interval.

The performance of the desiccant rotor may be influenced by distributions of temperature, humidity and amount of adsorbed water vapour in the desiccant rotor. In view of this, the distribution of adsorbed solar irradiation in the rotor is necessary to be analysed.

The mathematical model for numerical analysis is based on the following assumptions[1-2]:

(1) Air flow rate and pressure are constant during both adsorption and desorption processes.

(2) Considering only one-dimensional heat and mass transfer in the air flow in the air flow direction.

(3) Heat and mass transfer within the channel wall in the air flow direction can be ignored due to relatively low conductivity of adsorbent material.

(4) Temperature and amount of adsorbed moisture are uniform in each element wall, that is, adsorbent wall due to thin wall thickness.

(5) Amount of energy absorbed in each element in the rotor is evaluated from the distribution of absorbed solar irradiation[2].

The calculation procedure is as follows.

First, the desiccant rotor should be divided into many small elements in the direction of rotation, as shown in Fig. 5. 2 (a). The angle of rotor rotation $\theta$ corresponds to
the rotation time $t$. Both of the adsorption time $t_{ad}$ and desorption time $t_{de}$ in one rotation
are same, and 180s here. The cycle time $t_{cyc}$ is given by $t_{ad} + t_{de}$. For one rotating element, the calculation is conducted through the adsorption and desorption zones in the
rotation direction. Calculations are simultaneously performed for each sub element divided in the air flow direction as shown in Fig. 5. 2 (b). In this study, counter air flow is
considered between air flows in adsorption and desorption zones. Inlet air condition in
each zone is kept constant. The desiccant rotor thickness is separated into 10 sub elements with the interval of non-dimensional length 0. 1 as shown in Fig. 5. 2 (b). The
non-dimensional length $z/L$ stands for different meaning between adsorption process and
desorption process due to the counter air flow. In the adsorption process, $z/L = 0.1$ is
near the inlet of air flow path and $z/L = 0.9$ is near the outlet. While, in the desorption
process, vice versa.

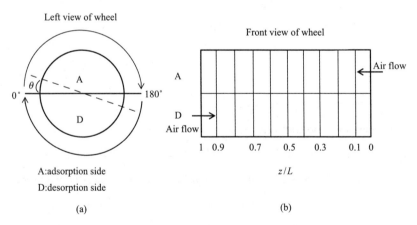

Fig. 5. 2 Desiccant rotor divided into many small elements in the direction of rotation and the air
flow direction in the calculation process

Secondly, the outlet air temperature and humidity of the element are calculated.
The adsorbed water in the desiccant wall is also calculated for adsorption and desorption
processes.

These calculations are continued until reaching a cyclic steady state. From the results of this simulation, the distribution of air and adsorbent bed temperatures, humidity and amount of adsorbed water vapour can be clarified.

## 5.1.2   Governing equations

From the assumptions, a one-dimensional model as shown in Fig.5.3 is used in this study. Governing equations are composed of mass and energy conservations both in the air flow path and the adsorbent wall, and adsorption rate. In these equations, it is necessary to consider heat and mass transfer coefficients between the air flow and adsorbent wall, adsorption equilibrium and heat of adsorption.

In addition, the input heat flux is given as a solar irradiation distribution received in the desiccant rotor. The absorption efficiency of irradiation energy is also introduced in a kind of distribution form, and the energy distribution equation and parameters obtained in Chapter 3 are used in the mathematical model.

Mass conservation in the air flow path is written as Eq. (5.1).

$$\rho_a \frac{\partial x}{\partial t} + \rho_a u_a \frac{\partial x}{\partial z} + \frac{m_{ad}}{a_a} = 0 \qquad (5.1)$$

where $\rho_a$ is the density of air flow [ kg ( DA )/m$^3$ ], $x$ is the absolute air humidity [ kg-water/kg( DA )], $t$ is the time (s), $z$ is the location in the direction of air flow (m), $u_a$ is the air flow velocity (m/s), $a_a$ is the equivalent thickness of air flow path (m) given with the ratio of the volume to the perimeter surface area of the air flow path, and $m_{ad}$ is the mass fluxfromthe adsorbent bed [ kg-water/( m$^2$ · s )].

Mass conservation in the adsorbentwall is written as Eq. (5.2).

$$m_{ad} = a_b \rho_b \frac{\partial q}{\partial t} \qquad (5.2)$$

where $a_b$ is a half of the rotor thickness ( m ), $\rho_b$ is the density of adsorbent wall ( kg-ads. /m$^3$ ), and $q$ is the amount of adsorbed water ( kg-water/kg-ads. ).

Adsorption rate isgiven as Eq. (5.3).

$$\frac{\partial q}{\partial t} = \frac{k_b}{a_b}(q^* - q) \qquad (5.3)$$

where $k_b$ is the equivalent mass transfer coefficient ( m/s ), and $q^*$ is the equilibrium adsorbed water in the adsorbent wall ( kg-water/kg-ads. ). Here, $k_b$was obtained from

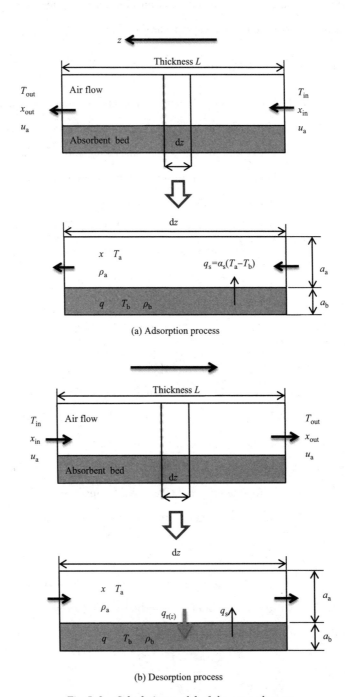

(a) Adsorption process

(b) Desorption process

Fig. 5. 3   Calculation model of the rotor element

the measurement of adsorption rate of water vapour combining with theoretical calculation for adsorbent wall sheet[2]. The equilibrium adsorbed water $q*$ is given by Eq. (5.4).

$$q^* = f\left(\frac{p_a}{p_{sat}}\right) = f(RH) \tag{5.4}$$

where $p_a$ is the partial pressure of water vapour in the air flow (Pa), and $p_{sat}$ is the saturate pressure of water vapour at the adsorbent wall temperature (Pa). The function in the right-hand side of Eq. (5.4) is different depending on the kinds of adsorption materials. The relation between the humidity $x$ and the partial pressure of water vapour $p_a$ is derived based on the assumption of the humid air treated as an ideal mixture gas of water vapour and dry air, and written as Eq. (5.5).

$$p_a = \frac{P}{x + 0.622}x \tag{5.5}$$

where, $P$ is the atmospheric pressure (Pa).

Energy conservation in the air flow path is written as Eq. (5.6).

$$\rho_a c_{pa} \frac{\partial T_a}{\partial t} + \rho_a u_a c_{pa} \frac{\partial T_a}{\partial z} + \frac{q_s}{a_a} = 0 \tag{5.6}$$

where $c_{pa}$ is the specific heat of air [J/(kg · K)], $T_a$ is the air temperature (K), $q_s$ is the heat flux (W/m$^2$) between the air flow and adsorbent wall given by the following equation, in which $\alpha_s$ is the heat transfer coefficient [W/(m$^2$ · K)].

$$q_s = \alpha_s(T_a - T_b) \tag{5.7}$$

Energy conservation in the adsorbent wall should be divided into two equations for calculationsin adsorption process and desorption process. In the adsorption process, the influence factors on the adsorption heat and heat transfer between the air flow and adsorbent wall should be taken into account. In the desorption process, input energy of solar irradiation should be considered. This energy input is the most important for regeneration of the desiccant rotor.

Energy conservation in the adsorbentwall during the adsorption process is written as Eq. (5.8).

$$\rho_b c_{pb} \frac{\partial T_b}{\partial t} = \rho_b \frac{\partial q}{\partial t} q_h + \frac{q_s}{a_b} \tag{5.8}$$

where $c_{pb}$ is the specific heat of adsorbent wall [J/(kg · K)], $T_b$ is the adsorbent wall temperature $(K)$, and $q_h$ is the heat of adsorption $(J/kg-water)$.

Energy conservation in the adsorbentwall during the desorption process is written as Eq. $(5.9)$.

$$\rho_b c_{pb} \frac{\partial T_b}{\partial t} = \rho_b \frac{\partial q}{\partial t} q_h + \frac{q_s}{a_b} + \frac{q_r}{a_b} \tag{5.9}$$

where $q_r$ is the normal component of heat flux $(W/m^2)$ from solar irradiation to the adsorbent wall surface. The details of heat flux $q_r$ is introduced as follows.

After concentrating by the Fresnel lens, the solar irradiation arrives in front of the desiccant rotor with different incident angles, as shown in Fig. 5.4. The incident angle increases with increasing the radial position of the desiccant rotor. The non-dimensional radius $r/R$ is used to express the relationship between the incident angle of solar ray and the radial location of the rotor $r$ $(m)$, in which $R$ is the radius of the desiccant rotor$(m)$.

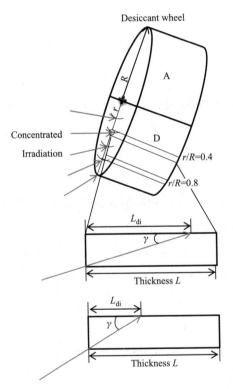

Fig. 5.4　Relationship between r/R and incident angle of solar irradiation (the arrow indicates the concentrated irradiation)

The incident angle increases lineally with an increase of $r/R$. Then, the length of direct irradiation impinging or heating zone $L_{di}$ in the inlet air flow portion decreases rapidly with the incident angle increases.

Fig. 5. 5 shows the distribution of solar heat flux energy in the rotor calculated by Eq. (5. 10). .

$$q_r(z) = \frac{q_{in}}{A_s L} \times \frac{L}{L_{di}} f_a(z) \qquad (5.10)$$

where $q_{in}$ is the input heat flux of concentrated solar irradiation at cross section of the rotor ($W/m^2$), $A_s$ is the volumetric surface area of the desiccant rotor ($m^2/m^3$), $L$ is the thickness or length of desiccant rotor (m), and $f_a(z)$ is the absorption efficiency of solar irradiation. The absorption efficiency $f_a(z)$ is introduced and determined based on the measuring data of the energy distribution in the desiccant rotorin Chapter 3.

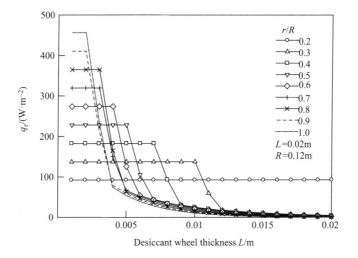

Fig. 5. 5　Distribution of solar heat flux $q_r(z)$

As seen in Fig. 5. 4 and Fig. 5. 5, when the non−dimensional radius $r/R$, that is, the incident angle of solar ray $\gamma$ is small, for example $r/R=0.2$, the absorption efficiency $f_a(z)$ is small at the cross edge part of the rotor, and then, keeps constant. Therefore, the solar ray irradiating in the centre portion of the rotor passes through the desiccant rotor pass with little absorption. While at the non−dimensional radius near or equal to 1, $r/R=0.9$ or $r/R=1$, the incident solar rays are much more and heat intensively the cross edge part of the desiccant rotor, that is, the direct heating zone. As shown in the Fig. 5. 5, the distribution characteristics in the desiccant rotor are clearly understood. This solar heat flux distribution is crucial for analysing the influence of input energy distribution on the desiccant rotor performance such as the distribution of humidity, air temperature, rotor temperature and amount of transferred water vapour in the adsorption and desorption processes.

## 5.1.3 Calculation condition

The structural specifications of the desiccant rotor used in the calculation are shown in Tab. 5. 1. Desiccant material of silica gel is filled into the rotor channel wall.

**Tab. 5. 1　Structural specifications of desiccant rotor**

| Item | Symbol | Unit | Value |
|---|---|---|---|
| Rotor diameter | $D_{rot}$ | m | 0. 24 |
| Boss diameter of the rotor | $D_{bos}$ | m | 0. 02 |
| Rotor thickness | $L$ | m | 0. 02 ~ 0. 2 |
| Mesh height of the desiccant rotor | $H$ | m | 0. 002 |
| Equivalent thickness of air flow path | $a_a$ | m | $0.32 \times 10^{-3}$ |
| A half of the bed thickness | $a_b$ | m | $0.1 \times 10^{-3}$ |
| Rotating angle at adsorption zone | $\theta_{ad}$ | rad | $\pi$ |
| Rotating angle at desorption zone | $\theta_{de}$ | rad | $\pi$ |
| Specific surface area of the rotor | $A_s$ | $m^2/m^3$ | 2400 |
| Length of direct heating zone | $L_{di}$ | m | $H/\tan\gamma$ |

The calculation conditions and parameters are shown in Tab. 5. 2. The subscripts ad stand for an adsorption process, de is a desorption process, in denotes inlet, cyc does cycle, a is air, and b is adsorbent wall. The inlet air temperatures, $T_{adin}$, and $T_{dein}$, and humidity, $x_{adin}$ and $x_{dein}$ are given as the boundary conditions in the calculation. Their values are set at a practical use of the desiccant system. Initial distributions of temperature, humidity and amount of adsorption water of the rotor are given arbitrarily. The overall heat and mass transfer coefficients, $\alpha_s$ and $k_b$ are quoted from the reference[3]. The size of Fresnel lens is 1m×0. 5m, the length of focus is 1. 2m, the concentration ratio of the solar irradiation of the Fresnel lens is $10$, the absorptivity of the rotor channel wall is $0.05$ and the mesh height of the desiccant rotor is $2$mm.

**Tab. 5.2 Calculation conditions and parameters**

| Item | Symbol | Unit | Value |
|---|---|---|---|
| Inlet air temperature of adsorption side | $T_{ad\ in}$ | K | 303 |
| Inlet air temperature of desorption side | $T_{de\ in}$ | K | 303 |
| Inlet air humidity of adsorption side | $x_{ad\ in}$ | kg/kg(DA) | 0.015 |
| Inlet air humidity of desorption side | $x_{de\ in}$ | kg/kg(DA) | 0.015 |
| Air flow velocity | $u_a$ | m/s | 0.2 |
| Cycle time of adsorption zone | $t_{ad}$ | s/rev | 180 |
| Cycle time in desorption zone | $t_{de}$ | s/rev | 180 |
| Cycle time | $t_{cyc}$ | s/rev | 360 |
| Rotor speed | $N$ | r/h | 10 |
| Air density | $\rho_a$ | kg/m$^3$ | 1.2 |
| Bed density | $\rho_b$ | kg/m$^3$ | 805 |
| Heat transfer coefficient | $\alpha_s$ | W/(m$^2 \cdot$ K) | 33 |
| Specific heat of air | $c_{pa}$ | J/(kg $\cdot$ K) | 1006 |
| Specific heat of adsorbent bed considering the adsorbed water | $c_{pb}$ | J/(kg $\cdot$ K) | 921+4180$q$ |
| Mass transfer coefficient | $k_b$ | m/s | $7\times10^{-5}$ |
| Absorptivity of the rotor | $a$ | — | 0.05 |
| Heat of adsorption | $q_h$ | kJ/kg | 2700 |
| Heat flux of concentrated solar irradiance (Base case) | $q_{in}$ | W/m$^2$ | 4000 |

The equilibrium adsorbed water $q^*$ is given as a function of relative humidity at the surface of adsorbent wall. The equilibrium adsorption was obtained from measurement. The measuring method is as follows. At first, the rotor was set in the constant partial vapour pressure atmosphere of air in the measurement room. The weight of rotor element was measured stepwise under different temperature conditions. The equilibrium adsorbed water $q^*$ was determined by dividing the amount of adsorbed water by the dry weight of the rotor including adsorbent at each relative humidity. The obtained relationship between the equilibrium adsorbed water $q^*$ [ kg – water/kg ] and relative humidity is

71

given by Eq. (5.11) and shown in Fig. 5.6.

$$q* = \frac{0.85(p_a/p_{sat})}{1 + 4.5(p_a/p_{sat})}$$ (5.11)

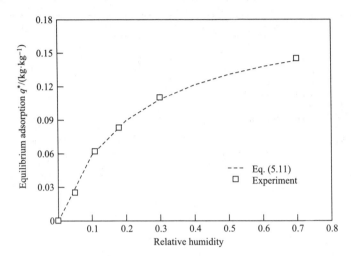

Fig. 5.6　Equilibrium adsorption

In the calculation, the number of mesh for the $z$ axis is set 30 and that for the $\theta$ axis is fixed at $3.6 \times 10^5$ ( corresponding to $\Delta t = 0.001 \text{s}$ ). The governing equations were solved by a finite difference method. First, $T$, $x$ and $q$ distributions are given based on the initial conditions. Then, the distributions of $T$, $x$ and $q$ of the next rotation angle element were calculated from the governing equations. This calculation process was continued from adsorption process to desorption process, until the values of $T$, $x$ and $q$ at $\theta$ $=0°$ and $\theta=360°$ were identical with each other within allowance value of $10^{-4}$.

## 5.1.4　Performance indices

In this paper, four indices are defined to assess the performance of the desiccant rotor; a difference between inlet and outlet of humidity, dehumidifying rate, local dehumidifying rate, and total dehumidifying rate.

The difference between inlet and outlet of humidity $\Delta x_{ad}(r/R)$ is written for adsorption process as Eq. (5.12), where $x_{adin}$ is the humidity at the inlet and $x_{adout}$ is the average outlet humidity during adsorption process. The average outlet humidity $x_{adout}$ varies at each radius location $r/R$ of the desiccant rotor.

$$\Delta x_{ad}(r/R) = x_{adin} - x_{adout}(r/R) \qquad (5.12)$$

The dehumidifying rate (namely, adsorption rate) $\Delta M_{ad}(r/R)$ is calculated with Eq. (5.13)[3]. It denotes the amount of adsorbed or desorbed water vapour per hour at each radius location $r/R$ of the desiccant rotor. Therefore, the dehumidifying rate is also regarded as humidifying rate.

$$\Delta M_{ad}(r/R) = \rho_a u_a \frac{\pi(D_{rot}^2 - D_{bos}^2)}{4}[x_{adin} - x_{adout}(r/R)] \times 3600 \qquad (5.13)$$

The local dehumidifying rate $\Delta M_{adloc}(r/R)$ is calculated by considering the volume fraction of $r/R$ annular section, as shown in Eq. (5.14).

$$\Delta M_{adloc}(r/R) = \Delta M_{ad}(r/R)\frac{V_r}{V_{rot}} \qquad (5.14)$$

where $V_r$ is the rotor volume of the annual section depending on the location of the rotor radius $r/R$ ($m^3$) and $V_{rot}$ is the total volume of the rotor ($m^3$). Fig. 5.7 shows the change of the local volume fraction $V_r/V_{rot}$ with the radial location.

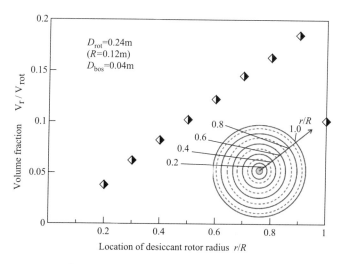

Fig. 5.7    Volume fraction of desiccant rotor

The total dehumidifying rate $\Delta M_{adtot}$ is the sum of local dehumidifying rates at all the locations of the desiccant rotor, as written as,

$$\Delta M_{adtot} = \sum_{r/R = 0.2}^{1} \Delta M_{adloc}(r/R) \tag{5.15}$$

where $r/R$ starts from $0.2$ except the rotor boss range.

## 5.2   Results and discussion

### 5.2.1   Comparison between experiment and simulation

In this section, the mathematical model will be examined using the result of measurement presented in the previous Chapter 4. In the measurement, the air flows during adsorption and desorption processes were regarded as parallel. For the comparison, the air flow situation was calculated in the simulation. The input energy of the solar light is 52.8W. The thickness of the rotor is 0.017m, and the radius of the rotor is 0.09m. The cycles times of adsorption process and desorption process are both 180s. The outlet humidity is the most important factor for predicting the performance of the desiccant rotor, so the outlet humidity is compared between simulation and experiment. The comparison of the outlet humidity is shown in Fig. 5.8 (a). The condition of inlet temperature is selected as 298K, the inlet relative humidity is 60%, and the air flow rate is 0.15m/s. In addition, the comparisons of the adsorption rate, that is, dehumidifying rate are shown in Figs. 5.8 (b), (c) and (d). The estimated measuring error of the humidity is ± 0.3 g/kg (DA) and the error of the dehumidifying rate is about ± 30%.

As shown in Fig. 5.8 (a), the outlet humidity predicted by simulation has similar trend to that of experiment, and the difference between simulation and experimental data is less than the measuring error. Also, the comparisons of the dehumidifying rate show similar trend in different conditions, and the difference between simulation and expement are within the measuring error range. Thus, it is confirmed that the present theoretical model is valid.

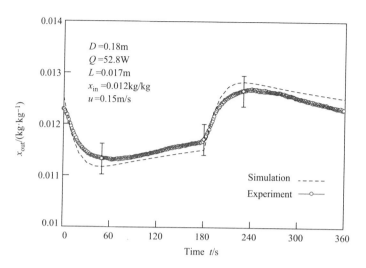

(a) Comparisons of the outlet humidity through the cycle

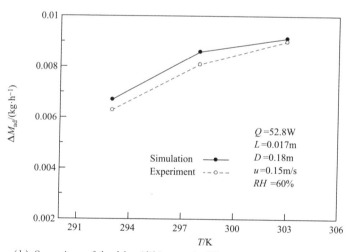

(b) Comparisons of the dehumidifying rate in different temperature conditions

Fig. 5. 8

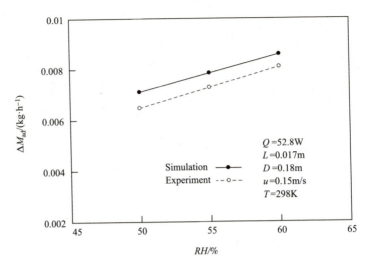

(c) Comparisons of the dehumidifying rate in different humidity conditions

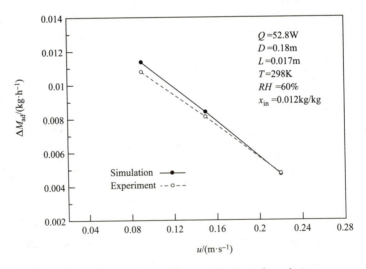

(d) Comparisons of the dehumidifying rate in different air flow velocity

Fig. 5. 8　Comparisons between experiments and calculations

## 5.2.2　Distributions of temperature, humidity, moisture removal and dehumidifying (adsorption) rate within desiccant rotor

In this section, simulation results for the calculation conditions in Tab. 5. 2, namely, the distributions of temperature, humidity and moisture removal in the air flow and

rotating directions are presented and discussed.

Fig. 5. 9 to Fig. 5. 12 show the distributions of bed temperature, air temperature, humidity and moisture removal (amount of adsorbed water vapour) on both adsorption side and desorption side during the periodically steady state. The angle from 0 to 180° in the left portion of figures denotes adsorption processes, while the angle from 180° to 360° in the right portion does desorption process, or regeneration process. The rotor thickness and rotation speed are 0. 02m and 10r/h, respectively.

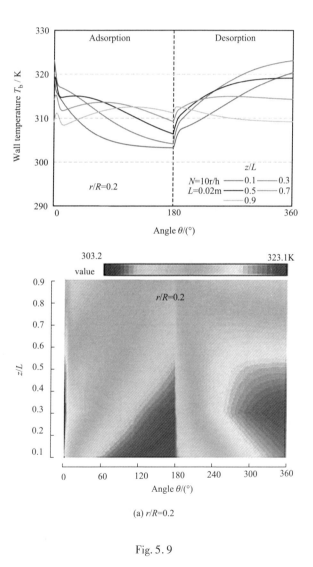

(a) $r/R=0.2$

Fig. 5. 9

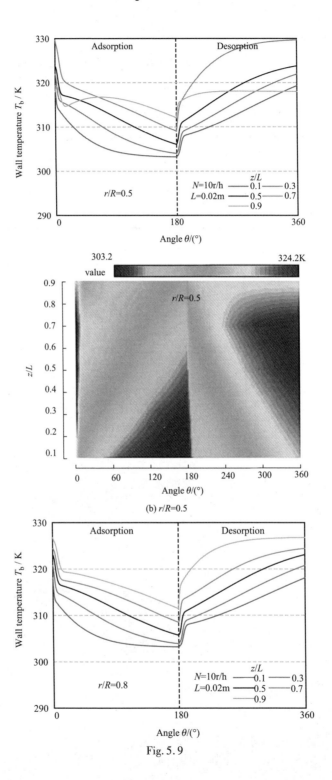

(b) $r/R$=0.5

Fig. 5.9

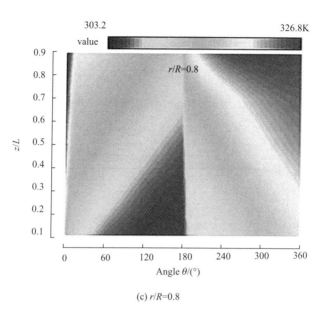

(c) $r/R$=0.8

Fig. 5. 9　Distribution of adsorbent wall temperature

In each figure, diagrams for the non-dimensional radius $r/R$ of 0. 2, 0. 5 and 0. 8 are shown. The $r/R$ = 0. 2, 0. 5 and 0. 8 represent solar ray arriving locations near the centre axis, at the middle part and near the outer edge of the desiccant rotor, respectively.

The distributions of adsorbent wall temperature in the rotor are shown in Fig. 5. 9. Generally, in the adsorption process, the wall temperature decreases with the rotation angle, while, in the desorption process, the wall temperature becomes higher with the angle. The difference among the different radial positions $r/R$ is marked at large $z/L$, where the adsorption heat is released greatly during adsorption process. The wall temperature distributions at the air flow position of 0. 1 and 0. 3 are not so different among the different $r/R$.

The wall temperature change at $z/L$ = 0. 7 and 0. 9 near the air flow inlet in the desorption process are evaluated as follows. At the non-dimensional radius $r/R$ = 0. 2 in Fig. 5. 9 (a), the large amount of solar rays passes through the air flow path to the rotor outside, and thus the wall temperature in the air flow inlet zone is maintained low. While, at $r/R$ of 0. 8 in Fig. 5. 9 (c), the concentrated irradiation intensively heats the inlet zone, and, therefore, the wall temperature in the inlet zone becomes significantly large more than that at $r/R = 0.2$.

As seen in Fig. 5.10, the distribution of the air temperature is entirely similar to the desiccant wall temperature distribution. But, the air temperature is generally smaller than the wall temperature. In the adsorption process, the air temperature at $z/L = 0.9$ becomes large, like the wall temperature in Fig. 5.9, due to the strong heat generation

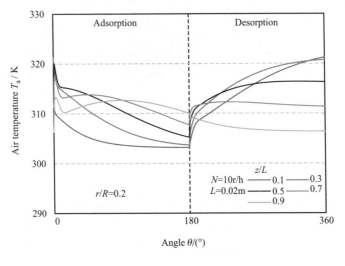

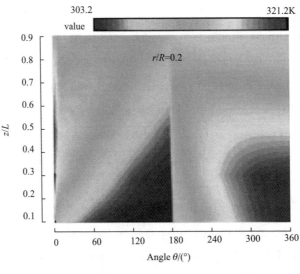

(a) $r/R=0.2$

Fig. 5.10

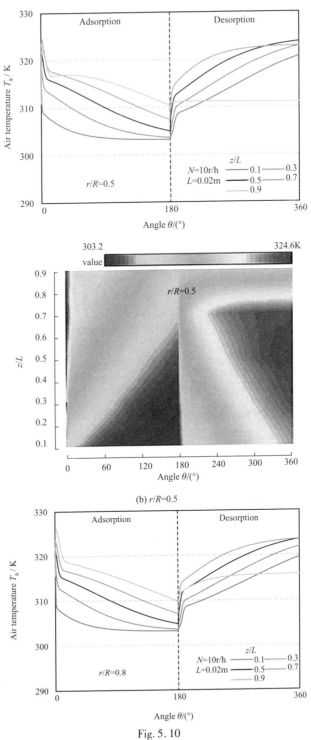

(b) $r/R$=0.5

Fig. 5. 10

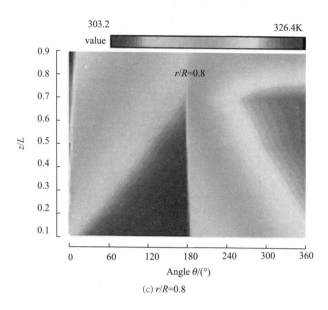

(c) $r/R$=0.8

Fig. 5. 10    Distribution of air temperature

within the wall.

In the desorption process, the solar irradiation heats the desiccant wall, and the desiccant wall releases the absorbed energy, namely, the sensible heat is transferred to the air from the adsorbent wall. Therefore, the air temperature increases with the wall temperature increase. Also, the air temperature rises in the rotating direction ($\theta$direction), markedly at $z/L$ = 0. 5 and $z/L$ = 0. 7. However, the air temperature near the air flow inlet, around $z/L$=0. 9, is kept almost constant, which is higher with increasing $r/R$. Thus, the inlet air cools the inlet zone very well.

The distributions of the air flow humidity are shown in Fig. 5. 11. In the adsorption process, as the non – dimensional radial position $r/R$ increases, the humidity near outlet, at $z/L$ = 0. 9, decreases markedly in the initial stage of the process, as seen in Fig. 5. 11 (c). It means that a lot of water vapour is adsorbed by the desiccant materials. Moreover, in the desorption process, the humidity increases entirely with increasing $r/R$, because the amount of water vapour is desorbed from adsorbent bed due to the high intensity of adsorbed solar irradiation. However, the humidity at inlet $z/L$ = 0. 9 does not change, and is kept almost constant, near the inlet humidity 0. 015kg/kg (DA).

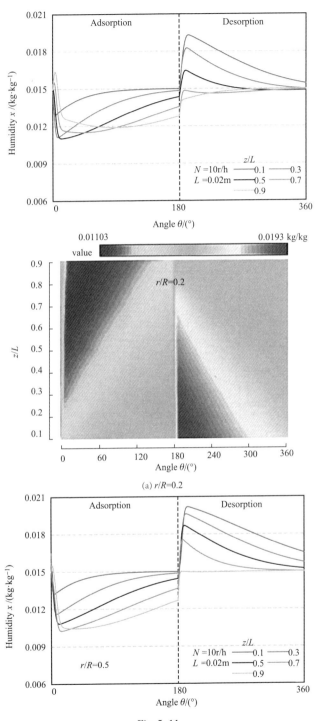

(a) $r/R$=0.2

Fig. 5. 11

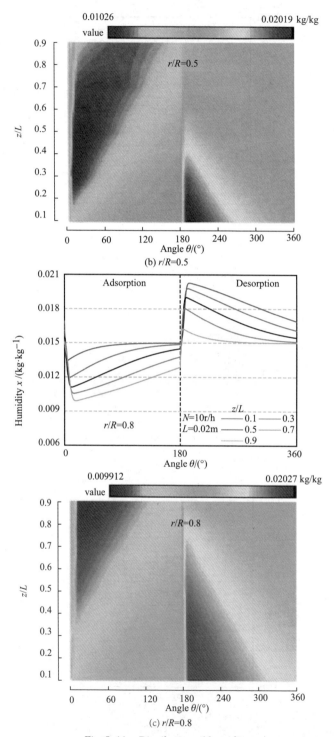

Fig. 5. 11   Distribution of humidity

The distributions of the amount of adsorbed water vapour $q$ ( moisture removal ) in the desiccant wall are shown in Fig. 5. 12. Generally, the change of the amount of adsorbed water vapour in each of the adsorption and desorption processes is large at $z/L = 0.1$ independently of $r/R$. While, at $z/L = 0.9$, the change in each process becomes small with decreasing $r/R$. At $r/R = 0.8$, the large change is seen on the whole of the flow path. The reason is the difference in the effect of solar irradiation on regeneration in the rotor, which directly determines the adsorption capacity in each location of rotor. So, it is confirmed that the direct heating method has high dehumidifying performance.

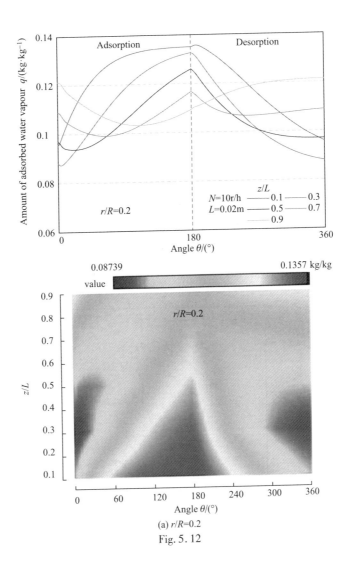

(a) $r/R$=0.2

Fig. 5. 12

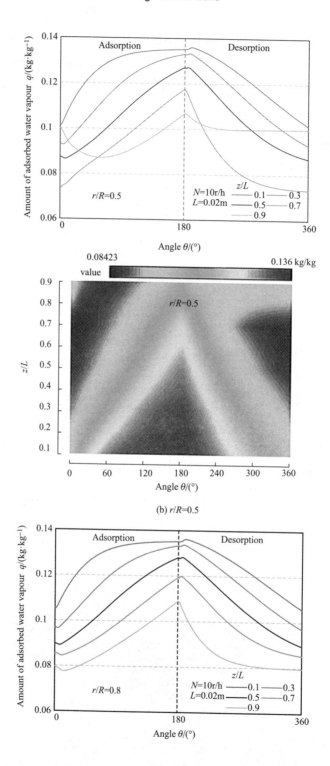

(b) $r/R$=0.5

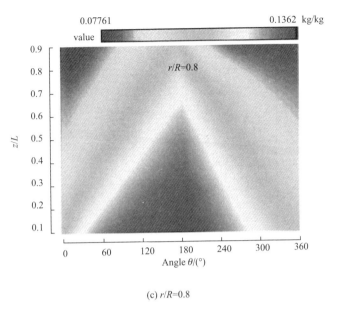

(c) $r/R$=0.8

Fig. 5. 12   Distribution of the amount of adsorbed water vapour

The distribution of calculated local adsorption rate $\Delta M_{adloc}$ is shown in Fig. 5. 13. The local adsorption rate $\Delta M_{adloc}$ increases as $r/R$ increases to 0.8. This is due to the volume fraction increases with increasing $r/R$. However, $\Delta M_{adloc}$ represents downward trendat $r/R$ = 0.9 to 1.0. This is the evidence that the solar irradiation onto the rotor inner wall is insufficient at $r/R$ = 0.9 or over with large irradiation incident angle. Besides, the volume fraction $V_r/V_{rot}$ at $r/R$= 1is almost half of that at $r/R$ = 0.9 as shown in Fig. 5. 7. Therefore, $\Delta M_{adloc}$ becomes small when $r/R$ reaches to 1.

Fig. 5. 14 shows the distribution of the difference of humidity $\Delta x_{ad}$ with the radial position $r/R$. The difference of humidity is small at $r/R$ less than 0.4 when the rotor thickness is 0.02m. It is attributed that, when the $r/R$ is small for the small rotor thickness, the concentrated solar irradiation passes through the desiccant rotor to outside. While, $\Delta x_{ad}$ shows almost constant in the range of $r/R$ = 0.4 to 0.9 for the rotor thickness of 0.04m and 0.06m. Besides, $\Delta x_{ad}$ falls when $r/R$ larger than 0.9 where the incident solar irradiation becomes small.

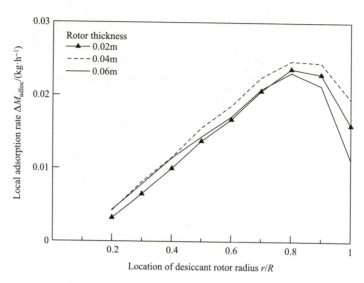

Fig. 5. 13    Influence of volume fraction on local adsorption rate

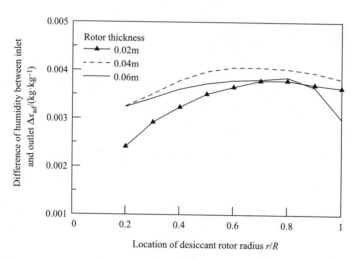

Fig. 5. 14    Distribution of humidity difference at each location of desiccant rotor radius

### 5.2.3    Influence of amount of input solar energy and rotor thickness on total adsorption rate

The total adsorption rate $\Delta M_{adtot}$ is influenced by the amount of input solar energy and rotor thickness as shown in Fig. 5. 15. The total adsorption rate $\Delta M_{adtot}$ increases with the input solar energy lineally. Input energy of 100W corresponds to solar heat flux at 200W/m² when the size of Fresnel lens is about 0.5m². Inputs of 200W and 400W correspond to solar heat fluxes of 400W/m² and 800W/m², respectively.

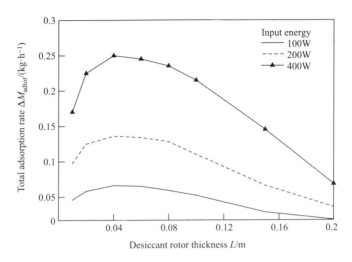

Fig. 5.15    Influence of amount of input energy on total adsorption rate

For the rotor thickness of around 0.04m, both local and total adsorption rates become maximum as shown in Fig. 5.13 and Fig. 5.15. When less than 0.04m, the $\Delta M_{adtot}$ decrease because of short contacting time between air flow and adsorbent. While, for $L$ larger 0.08 m, $\Delta M_{adtot}$ reduces gradually with increasing $L$, and this is caused by the re-adsorption phenomenon in desorption process of desorbed water vapour due to long contacting time between air flow and adsorbent. Fig. 5.16 shows the comparison of air humidity and desiccant wall temperature distributions between $L = 0.02$m and $L = 0.2$m. Both in adsorption and desorption processes, the humidity and wall temperature change become small markedly when $L=0.2$m, compared to $L = 0.02$m. In this manner, there-adsorption phenomenon is confirmed for the long rotor thickness.

## 5.2.4    Comparison between direct and indirect heating types

For heating a desiccant rotor, indirect methods such as air heating or water heating with solar collectors have been focused in most studies[4]. However, in indirect methods, the regeneration temperature cannot become high due to low thermal efficiency of

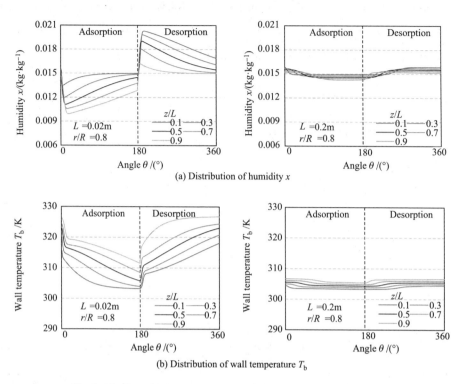

(a) Distribution of humidity $x$

(b) Distribution of wall temperature $T_b$

Fig. 5. 16　Distribution comparison between $L=0.02$m and $L=0.2$m

collectors. By Sopian K et al.[5], the thermal efficiency is generally $60\% \sim 70\%$. Here, the average value $65\%$ is selected as the thermal efficiency of indirect type. To clarify the difference in heating type, the calculation also performed for the indirect type using the present mathematical model in which, for regeneration of the desiccant wall during desorption process, a hot air inflow is given, in place of solar irradiation. In the calculation, the case of ideal thermal efficiency $100\%$ is also considered.

Fig. 5. 17 shows the comparison of the total adsorption rate $\Delta M_{adtot}$ between the present direct heating type and indirect heating type. Both the results of direct type and indirect type with $100\%$ efficiency are recognized to be almost the same. While, compared with the indirect type of thermal efficiency $65\%$, the present heating type has $40\%$ higher performance.

In some cases, it is important to increase the difference between inlet and outlet of humidity of the rotor. The average of the humidity difference between inlet and outlet of the desiccant rotor is shown in Fig. 5. 18. The humidity difference of indirect type with 65% thermal efficiency is smaller than that of the direct type.

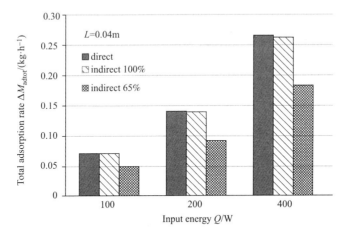

Fig. 5. 17　Influence of heating type on total adsorption rate

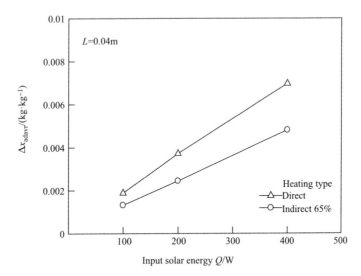

Fig. 5. 18　Influence of heating type on humidity difference between inlet and outlet

Both Fig. 5. 17 and Fig. 5. 18 clarify that the dehumidification ability of the direct heating type is stronger than that of the indirect one.

## 5.2.5　Influence of heat transfer coefficient between bed and air

The heat transfer coefficient $\alpha_s$ between adsorbent wall and air flow is a significant influence factor for the total adsorption rate $\Delta M_{adtot}$ of the desiccant rotor. As the coefficient increases, the efficiency of the heat transfer improves, and then the adsorption rate increases[6]. Fig. 5.19 shows the effect of the heat transfer coefficient on the performance of direct and indirect types. With increasing of $\alpha_s$ tenfold from $3.3\mathrm{W}/(\mathrm{m}^2 \cdot \mathrm{K})$ to $33\mathrm{W}/(\mathrm{m}^2 \cdot \mathrm{K})$, both heating types show the increasing of $\Delta M_{adtot}$. The total adsorption rate $\Delta M_{adtot}$ of the indirect type increases by 72 %, which is larger than three times 23% of the direct type. So, indirect heating type is much affected by $\alpha_s$ because the rotor is heated by hot air, not by direct heating. It is expected that the direct type does not require much air flow rate to increase heat transfer coefficient, and, consequently, it can reduce a pressure drop of air flow through the rotor.

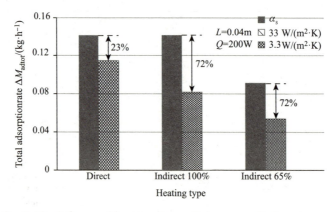

Fig. 5.19　Influence of heat transfer coefficient on total adsorption rate

## 5.2.6　Influence of air flow velocity(contacting time)

The air flow velocity in the rotor also influences the desiccant rotor performance. For a fixed desiccant rotor thickness, as the air flow velocity increases, the contacting time of air flow with the desiccant wall becomes smaller. The effect of the air flow velocity on the total adsorption rate is shown in Fig. 5.20(the first row of X-coordinate is air flow velocity, the second row is contacting time). The total adsorption rate decreases with the increase of air flow speed. The effect of the air flow speed is greater for the direct type compared with the indirect ones. In case of high velocity, namely, significantly short contact time, for example 0.02s in Fig. 5.20, the performance of the direct heating type becomes worse. However, this condition is extremely short contact time for usual desiccant rotor system, that is, an exceptional case.

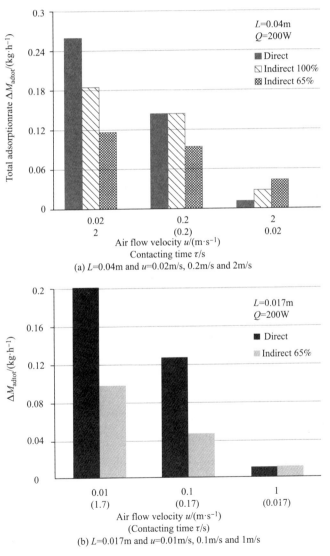

Fig. 5.20    Influence of air flow speed on total adsorption rate

## 5.3　Conclusions

In thischapter, the theoretical model for the desiccant rotor heated directly by concentrated solar irradiation using Fresnel lens was proposed considering the distribution of solar irradiation in the rotor. By comparing between simulation and experiment, the validity of the theoretical model was confirmed. Based on the numerical simulation, the influences of operation parameters and desiccant rotor characteristics were clarified. The obtained results are as follows.

(1)The distributions of desiccant wall temperature, air temperature, humidity, a-mount of adsorbed water vapour and adsorption rate were influenced by the distribution of concentrated solar irradiation in the rotor. In the desorption process, the wall temperature becomes higher with the rotation angle. The difference among the different radial positions is marked at large $z/L$, and this is because the intensity of solar irradiation is very high near the inlet zone of desiccant rotor. The distribution of air temperature showed entirely a similar trend to the wall temperature. The humidity in the initial stage of adsorption process decreased markedly at outlet zone as $r/R$ increased. The change of the amount of adsorbed water vapour in the desorption process was generally large, except for the portion of small radius near the out let zone. It was confirmed that the concentrated solar irradiation was effective to heat the desiccant rotor entirely.

(2)Generally, the local adsorption rate and difference of humidity increased with increasing the radial position, although they got smaller for $r/R$ more than 0.8 where the solar irradiation not lighting on the rotor wall.

(3)The adsorption rate increased linearly with the increase of the input solar energy.

(4)There was a suitable, that is, optimum desiccant rotor thickness in which the adsorption rate reached a maximum.

(5)The total adsorption rate of direct type was generally 40% higher than the indirect type with thermal efficiency of collector 65%.

(6)With the increase of heat transfer coefficient between air and wall, the total adsorption rate became large. In the direct heating type, the effect of heat transfer coefficient was small compared to the indirect type.

(7)The total adsorption rate decreased with the increase of the air flow velocity. Influence of the air flow velocity was much more significant in the direct type than the indirect type.

# References

[1]    HAMAMOTO Y, TRAN T N , AKISAWA A, et al. Experimental and numerical study of desiccant rotor with direct heating regeneration by solar energy[C]. Proceedings of the 6th ASME – JSME Thermal Engineering Joint Conference, TED-AJ03-273, 2003.

[2]    HAMAMOTO Y, MORI H. Influence of energy distribution in a desiccant rotor with direct heating process by concentrated solar irradiation on the dehumidification performance[C]. Proceedings of the International Sorption Heat Pump Conference, 2008, AB-057, 2008.

[3]    HAMAMOTO Y, MORI H. A model of absorbed energy distribution and numerical simulation in a desiccant rotor regenerated by concentrated solar irradiance [J]. Proceedings of Renewable Energy, 2010, 33, 1-4.

[4]    RUIVO C R , GOLDSWORTHY M ,INTINI M. Interpolation methods to predict the influence of inlet airflow states on desiccant wheel performance at low regeneration temperature[J]. Energy, 2014, 68, 765-772.

[5]    SOPIAN K, ALGHOUL M A, ALFEGI E M, et al. Evaluation of thermal efficiency of double-pass solar collector with porous-nonporous media[J]. Renewable Energy,2009, 34, 640-645.

[6]    HAMAMOTO Y, MORI H, GODO M, et al. Overall heat and mass transfer coefficient of water vapour adsorption-2nd report: transfer coefficient for adsorbent rotor blocks[J]. Transactions of the Japan Society of Refrigerating and Air Conditioning Engineers, 2007, 24, 473-484.